Anna Bastian

Diversity in the Cuban bat species, Macrotus waterhousii

Anna Bastian

Diversity in the Cuban bat species, Macrotus waterhousii

Relevance of variation in echolocation, foraging behaviour, social interactions and genetic markers in two morphotypes

Südwestdeutscher Verlag für Hochschulschriften

Impressum / Imprint
Bibliografische Information der Deutschen Nationalbibliothek: Die Deutsche Nationalbibliothek verzeichnet diese Publikation in der Deutschen Nationalbibliografie; detaillierte bibliografische Daten sind im Internet über http://dnb.d-nb.de abrufbar.
Alle in diesem Buch genannten Marken und Produktnamen unterliegen warenzeichen-, marken- oder patentrechtlichem Schutz bzw. sind Warenzeichen oder eingetragene Warenzeichen der jeweiligen Inhaber. Die Wiedergabe von Marken, Produktnamen, Gebrauchsnamen, Handelsnamen, Warenbezeichnungen u.s.w. in diesem Werk berechtigt auch ohne besondere Kennzeichnung nicht zu der Annahme, dass solche Namen im Sinne der Warenzeichen- und Markenschutzgesetzgebung als frei zu betrachten wären und daher von jedermann benutzt werden dürften.

Bibliographic information published by the Deutsche Nationalbibliothek: The Deutsche Nationalbibliothek lists this publication in the Deutsche Nationalbibliografie; detailed bibliographic data are available in the Internet at http://dnb.d-nb.de.
Any brand names and product names mentioned in this book are subject to trademark, brand or patent protection and are trademarks or registered trademarks of their respective holders. The use of brand names, product names, common names, trade names, product descriptions etc. even without a particular marking in this works is in no way to be construed to mean that such names may be regarded as unrestricted in respect of trademark and brand protection legislation and could thus be used by anyone.

Coverbild / Cover image: www.ingimage.com

Verlag / Publisher:
Südwestdeutscher Verlag für Hochschulschriften
ist ein Imprint der / is a trademark of
AV Akademikerverlag GmbH & Co. KG
Heinrich-Böcking-Str. 6-8, 66121 Saarbrücken, Deutschland / Germany
Email: info@svh-verlag.de

Herstellung: siehe letzte Seite /
Printed at: see last page
ISBN: 978-3-8381-3559-5

Zugl. / Approved by: Hannover, Stiftung Tierärtzliche Hochschule Hannover, Diss., 2012

Copyright © 2012 AV Akademikerverlag GmbH & Co. KG
Alle Rechte vorbehalten. / All rights reserved. Saarbrücken 2012

Table of contents

ABSTRACT I
ZUSAMMENFASSUNG II

1. INTRODUCTION 1

2. MATERIAL AND METHODS 11

2.1 DATA BASE 11
2.2 GENERAL METHODS 12
2.2.1 AUDIO RECORDINGS AND ANALYSES 12
2.2.2 VIDEO RECORDINGS AND ANALYSES 15
2.3 EXPERIMENT SPECIFIC METHODS 16
2.3.1 FIELD STUDIES 16
2.3.1.1 Morphometric measurements 16
2.3.1.2 Orientation pulses in a standardized situation 17
2.3.2 BEHAVIOURAL STUDIES IN THE LAB 18
2.3.2.1 Laboratory facilities: Flight room and experimental room with experimental cage 18
2.3.2.2 Training of the bats 20
2.3.2.3 Passive listening 21
2.3.2.4 Prey capture success under two different clutter conditions 24
2.3.2.5 Manoeuvrability 26
2.3.2.6 Social interactions 27
2.3.2.6.1 Allogrooming partners 28
2.3.2.6.2 Social calls 28
2.3.2.7 Spatial allocation 28
2.3.3 STATISTICS FOR FIELD STUDIES AND STUDIES ON BEHAVIOUR IN THE LAB ... 30
2.4 GENETIC VARIABILITY 32
2.4.1 GENE SEQUENCING PROCEDURES 32
2.4.1.1 Sequence analyses 34

3. RESULTS 38

3.1 GENERAL OBSERVATIONS ON CUBAN M. WATERHOUSII 38
3.2 MORPHOLOGY 40
3.2.1 DESCRIPTION AND COMPARISON OF MORPHOMETRICAL DATA OF CUBAN M. WATERHOUSII 40
3.2.1.1 Fur coloration 40
3.2.1.2 Multiparametrical analyses of morphometry 42
3.2.1.2.1 Summary 50
3.2.2 MANOEUVRABILITY 51
3.2.2.1 Summary 54
3.2.3 SUMMARY OF MORPHOLOGY 54
3.3 ACOUSTIC ORIENTATION AND FORAGING BEHAVIOUR 55
3.3.1 DESCRIPTION AND COMPARISON OF ORIENTATION PULSES OF CUBAN M. WATERHOUSII 55
3.3.1.1 Multiparametrical analysis of echolocation pulses 56
3.3.1.1.1 Summary 65
3.3.2 PASSIVE LISTENING 66
3.3.2.1 Summary 69
3.3.3 PREY CAPTURE SUCCESS UNDER DIFFERENT CLUTTER CONDITIONS 69
3.3.3.1 Summary 78
3.3.4 SUMMARY OF ACOUSTIC ORIENTATION AND FORAGING BEHAVIOUR 78
3.4 SOCIAL INTERACTIONS 79
3.4.1 SPATIAL ALLOCATION 79
3.4.2 ALLOGROOMING PARTNERS 82
3.4.3 SOCIAL CALLS 83
3.4.4 SUMMARY 85
3.5 GENETIC VARIABILITY 86
3.5.1 SUMMARY 97

4. DISCUSSION 98

5. APPENDIX 109

5.1	Taxonomic classification of Macrotus	109
5.2	Distances between the sample sites	111
5.3	Emission of the fundamental in echolocation pulses	114
5.4	Photos of reproductive status of M. waterhousii	115
5.5	List of individuals kept in the lab	116
5.6	Photos of defined areas used in the experiment on spatial allocation	118
5.7	Analysed sessions in the experiment on passive listening	120
5.8	Descriptive statistics for morphometric measurements	121
5.9	Correlation between morphometrical parameters	123
5.10	Multiparametrical comparison of the four sample sites based on morphology	124
5.11	Classification of individuals based on morphology	128
5.12	Flight speed of bats in the experiment on manoeuvrability	129
5.13	Descriptive statistics of echolocation pulse measurements	132
5.14	Correlations between the acoustical parameters	135
5.15	Multiparametrical comparisons of the four sample sites based on echolocation	136
5.16	Additional results of the experiment on prey capture success in clutter	140
5.17	Presence of bats at defined areas in the facility	152
5.18	Allogrooming events of bat dyads in the facility	153
5.19	Additional descriptive results of sequence analyses	154

6. LIST OF REFERENCES 161

List of figures

FIGURE 1: DISTRIBUTION AREA OF THE GENUS *MACROTUS*. 3
FIGURE 2: LOCATION OF CAVES SAMPLED IN THIS STUDY. 11
FIGURE 3: SKETCH OF THE FACILITY. 19
FIGURE 4: RUSTLING NOISE PROFILES OF THE THREE PRESENTED FEEDING DISHES. 22
FIGURE 5: ECHO CHARACTERISTICS OF THE THREE CLUTTER TYPES. 25
FIGURE 6: MORPHOMETRIC PARAMETERS PLOTTED AS A FUNCTION OF LONGITUDE. 43
FIGURE 7: MORPHOMETRIC PARAMETERS OF BATS FROM THE FOUR SAMPLE SITES. 45
FIGURE 8: PLOT OF FIRST TWO CANONICAL FUNCTIONS OF MULTIPLE MORPHOLOGICAL PARAMETERS. 50
FIGURE 9: FLIGHT SPEED UNDER THREE EXPERIMENTAL CONDITIONS. 52
FIGURE 10: A TYPICAL ECHOLOCATION PULSE EMITTED BY CUBAN *M. WATERHOUSII*. 56
FIGURE 11: ACOUSTIC PARAMETERS OF BATS FROM FOUR SAMPLE SITES. 59
FIGURE 12: PLOT OF FIRST TWO CANONICAL FUNCTIONS OF MULTIPLE ACOUSTIC PARAMETERS. 65
FIGURE 13: NUMBER OF APPROACHES OF EACH INDIVIDUAL TO THE FEEDING DISHES. 67
FIGURE 14: NUMBER OF LANDINGS FOR EACH INDIVIDUAL ON THE FEEDING DISHES. 68
FIGURE 15: BOXPLOTS SHOWING APPROACHES OF BATS TO EXPERIMENTAL FIELDS. 71
FIGURE 16: BOXPLOTS SHOWING LANDINGS OF BATS TO EXPERIMENTAL FIELDS. 74
FIGURE 17: PRESENCE OF BATS AT DEFINED AREAS INSIDE THE FLIGHT ROOM. 80
FIGURE 18: SONAGRAMS OF TYPICAL CALLS OF THREE COMMONLY EMITTED SOCIAL CALL TYPES. 85
FIGURE 19: MINIMUM SPANNING HAPLOTYPE NETWORK OF THE D-LOOP REGION AND CYT-B GENE. 91
FIGURE 20: PHYLOGRAM OF D-LOOP AND CYT-B OF SEQUENCES. 92
FIGURE 21: VARIABILITY PATTERN OF MORPHOLOGY. 100
FIGURE 22: VARIABILITY PATTERN OF ECHOLOCATION PULSE PARAMETERS. 102
FIGURE 23: VARIABILITY PATTERN OF COMBINED PHYLOGENY AND HAPLOTYPE RESULTS OF BOTH MARKERS. 104
FIGURE 24: TWO-COLONISATION SCENARIO INCLUDING THE RESULTS OF THIS STUDY. 108
FIGURE 25: LACTATING MACROTUS FEMALE. 115
FIGURE 26: BOLD-HEADED MALE *MACROTUS*. 115
FIGURE 27: PHOTOS OF THE DEFINED AREAS INSIDE THE FLIGHT ROOM. 118

List of tables

TABLE 1: LIST OF ACOUSTIC PARAMETERS MEASURED IN ECHOLOCATION PULSES. 13
TABLE 2: LIST OF MORPHOMETRIC PARAMETERS TAKEN FOR EACH INDIVIDUAL. 16
TABLE 3: PARAMETERS USED FOR THE ANALYSIS OF THE EXPERIMENT ON PASSIVE LISTENING. 23
TABLE 4: DEFINED AREAS IN THE FLIGHT ROOM WHERE BATS WERE PERMANENTLY HOUSED. 29
TABLE 5: STANDARD MOLECULAR DIVERSITY AND POPULATION GENETICS MEASUREMENTS. 37
TABLE 6: COLORATION TYPES AND THEIR FREQUENCIES AT SAMPLE SITES. 41
TABLE 7: SEXUAL DIMORPHISM OF FUR COLORATION. 42
TABLE 8: RESULTS OF MULTIPARAMETRICAL PAIR WISE COMPARISONS FOR EACH MORPHOLOHICAL PARAMETER. 48
TABLE 9: NUMBER OF COLLISIONS OF THE BATS WITH THE SECOND OBSTACLE IN THE TWO-OBSTACLE FLIGHT COURSE. 53
TABLE 10: OVERVIEW OF EMISSION OF THE FUNDAMENTAL IN ECHOLOCATION PULSES. 58
TABLE 11: RESULTS OF MULTIPARAMETRICAL PAIR WISE COMPARISONS FOR EACH ACOUSTIC PARAMETER. 61
TABLE 12: CLASSIFICATION RESULTS OF THE DISCRIMINANT ANALYSIS. 63
TABLE 13: APPROACHES OF BATS TO THE REWARD OR NON-REWARD FIELDS IN DIFFERENT CLUTTER CONDITIONS. 73
TABLE 14: LANDINGS OF BATS IN THE REWARD OR IN THE NON-REWARD FIELDS IN DIFFERENT CLUTTER CONDITIONS. . 76
TABLE 15: OBSERVED FREQUENCIES AND TEST VALUES FOR EACH PREDEFINED AREA IN THE FACILITY PER BAT. 81
TABLE 16: SUMMARIZED RESULTS OF ALLOGROOMING DYADS. 82
TABLE 17: SUMMARY STATISTICS OF CYTOCHROM-B SEQUENCE DATA. 86
TABLE 18. ESTIMATES OF EVOLUTIONARY DIVERGENCE BETWEEN CYT-B SEQUENCES. 88
TABLE 19: BETWEEN SAMPLE SITE GENETIC PAIRWISE DIFFERENCES OF CYT-B SEQUENCES. 90
TABLE 20: RESULTS OF AMOVA OF CYT-B SEQUENCES. 90
TABLE 21: SUMMARY STATISTICS OF D-LOOP SEQUENCE DATA. 93
TABLE 22. ESTIMATES OF EVOLUTIONARY DIVERGENCE BETWEEN D-LOOP SEQUENCES. 95
TABLE 23: BETWEEN SAMPLE SITE GENETIC PAIRWISE DIFFERENCES OF D-LOOP SEQUENCES. 96
TABLE 24: RESULTS OF AMOVA OF D-LOOP3 SEQUENCES. 96
TABLE 25: OVERVIEW OF TAXONOMIC CLASSIFICATIONS WITHIN THE GENUS *MACROTUS* 109
TABLE 26: DISTANCES BETWEEN THE FOUR SAMPLE AREAS IN KM 111
TABLE 27: CROSS-TABLE OF DISTANCES BETWEEN THE CAVES IN KM. 111
TABLE 28: OVERVIEW OF NUMBERS OF PULSES WITH A PROMINENT 2ND HARMONIC. 114
TABLE 29: LIST OF BATS KEPT IN A FACILITY AT THE UNIVERSITY OF HAVANA IN THE YEARS 2008, 2009/2010. 116
TABLE 30: OVERVIEW OF ANALYZED SESSIONS IN THE EXPERIMENT ON PASSIVE LISTENING. 120

TABLE 31: DESCRIPTIVE STATISTICS OF MORPHOMETRIC PARAMETERS MEASURED. 121
TABLE 32: SPEARMAN CORRELATION COEFFICIENTS BETWEEN THE MORPHOLOGICAL PARAMETERS. 123
TABLE 33: RESULTS OF MULTIPLE GROUP COMPARISON BY MORPHOLOGICAL PARAMETERS. 124
TABLE 34: SIGNIFICANT RESULTS OF MULTIPLE PAIRWISE COMPARISONS OF MORPHOLOGICAL PARAMETERS. 125
TABLE 35: RESULTS OF MANOVA POST HOC TESTS OF MORPHOMETRICAL PARAMETERS. 126
TABLE 36: CLASSIFICATION RESULTS OF THE DISCRIMINANT ANALYSIS ON MORPHOLOGICAL CHARACTERISTICS. 128
TABLE 37: FLIGHT SPEED OF 14 INDIVIDUALS IN THREE EXPERIMENTAL CONDITIONS. 129
TABLE 38: DESCRIPTIVE STATISTICS FOR ECHOLOCATION PULSES. 132
TABLE 39: CORRELATION COEFFICIENTS BETWEEN THE MORPHOLOGICAL MEASURES AND THE ACOUSTIC MEASURES. 135
TABLE 40: RESULTS OF MULTIPLE GROUP COMPARISON BY MEDIAN CALL PARAMETERS 136
TABLE 41: SIGNIFICANT RESULTS OF MULTIPLE PAIRWISE COMPARISONS OF ACOUSTIC PARAMETERS. 137
TABLE 42: RESULTS OF MANOVA POST HOC TEST OF ACOUSTIC PARAMETERS. 139
TABLE 43: RESULTS OF CHI^2 STATISTICS FOR APPROACHES TO DISHES WITHOUT SUBSEQUENT LANDING. 140
TABLE 44: RESULTS OF CHI^2 STATISTICS FOR APPROACHES TO DISHES WITH SUBSEQUENT LANDING. 141
TABLE 45: FREQUENCIES OF FIRST APPROACHES TO PREVIOUS REWARD FIELD. 142
TABLE 46: NUMBER OF HOVERING EVENTS PRIOR LANDINGS. 143
TABLE 47: NUMBER OF APPROACHES TO REWARD FIELD AND NON-REWARD FIELD IN THREE CLUTTER CONDITIONS... 144
TABLE 48: NUMBER OF LANDINGS IN THE REWARD AND NON-REWARD FIELD IN THREE CLUTTER CONDITIONS.......... 148
TABLE 49: CHI^2 VALUES OF OBSERVED NUMBERS AT WHICH BATS WERE PRESENT AT EACH AREA IN THE ROOM. 152
TABLE 50: NUMBER OF ALLOGROOMING EVENTS FOR EACH DYAD. 153
TABLE 51: DIAGNOSTIC SITES OF CYT-B SEQUENCES. 154
TABLE 52: LIST OF HAPLOTYPES IN THE CYT-B DATA SET. 155
TABLE 53: DIAGNOSTIC SITES OF D-LOOP SEQUENCES. 157
TABLE 54: LIST OF HAPLOTYPES IN THE D-LOOP DATA SET. 158

Author: Anna Bastian. Title: "Acoustic basis for biodiversity: Variability in echolocation, foraging behaviour, social interactions and genetic markers in two morphotypes of the Cuban bat, *Macrotus waterhousii*"

Abstract

Previous studies indicated a multi-species complex in the neotropical bat species *Macrotus waterhousii*. Two of the proposed seven morphotypes occur on Cuba, a small morphotype in the western part and a large morphotype in the eastern part with no intermediate forms in the central region. This pattern can be the result of two colonisation events from the Central American mainland, or due to a cline. If the described lack of hybrids is indeed existent, isolation barriers might be responsible for this separation. It was the aim of my study, to provide a profound assessment of Cuban *M. waterhousii*´s variability by a comprehensive approach. Individuals sampled in the field at four geographical regions on Cuba were compared with regard to differences in morphology, echolocation pulse design, and genetic divergence. Furthermore, experiments on manoeuvrability, prey capture success in a cluttered environment and social interactions were conducted with a group of individuals from three sample sites housed in a facility to reveal indications of isolating barriers. There were no hints to adaptations into microhabitats in manoeuvrability and in echolocation pulses referring to ecological barriers or to reduced interactions between individuals from different sample sites referring to ethological barriers. Thus, differences between sample sites were due to genetic drift rather than positive selection. In contrast to previous studies, no distinct morphotypes were obvious. The pattern of variability is more complex in all approaches and is best explained by a two-colonisation scenario. A western founder population from Yucatán and an eastern founder population which reached Cuba via Jamaica and Hispaniola moved towards the centre of the Cuba, both further diverging while moving. They came in a second contact and hybridized in the centre of the island, then locally developing distinct sets of characteristics as revealed in this study.

Autorin: Anna Bastian. Titel: "Acoustic basis for biodiversity: Variability in echolocation, foraging behaviour, social interactions and genetic markers in two morphotypes of the Cuban bat, *Macrotus waterhousii*"

Zusammenfassung

Vorangegangenen Studien nach stellt die Art *M. waterhousii* einen Artkomplex dar. Sie ist über das zentralamerikanische Festland und die Inseln der Karibik verbreitet, wo sie lokale Morphotypen ausbildet, deren taxonomischer Status bislang ungeklärt ist. Es wird angenommen, dass es auf Kuba zwei distinkte Morphotypen gibt; ein kleiner Typus der im Westen der Insel vorkommt und ein größerer Morphotyp der im Osten verbreitet ist. Es wurden keine Hybridformen im Zentrum der Insel gefunden. Dieses Verbreitungsmuster könnte entweder auf Grund zweier Besiedlungswellen entstanden sein oder durch eine Kline. Kommt es in einer Begegnungszone zweier Formen nicht zu Hybriden, dann sind Isolationsbarrieren zu vermuten. Ziel dieser Studie ist es, die Variabilität der kubanischen Population von *M. waterhousii* zu erfassen und zu bewerten. Dazu wurde ein Ansatz gewählt, der die Morphologie, Echoortung und genetische Divergenz zwischen Gruppen von Individuen aus vier Regionen der Insel untersucht und vergleicht. Zudem wurden Experimente zur Manövrierfähigkeit, zum Beutefangerfolg in echoreicher Umgebung und zu sozialen Interaktionen an einer in Gefangenschaft lebenden Gruppe aus Individuen aus drei Regionen durchgeführt, um Hinweise auf Isolationsbarrieren zu untersuchen. Die Resultate zeigten keine Hinweise auf Isolationsbarrieren, weder auf ökologische Barriereformen durch Einnischungen in Mikrohabitate in Manövrierfähigkeit und Echoortung beim Beuteerwerb, noch auf ethologische Barrieren in den Analysen der sozialen Interaktionen. Zudem konnte die Existenz von zwei distinkten Morphotypen auf Kuba nicht bestätigt werden. Die Resultate der Vergleiche zwischen den vier Regionen machen wahrscheinlich, dass *M. waterhousii* in zwei Besiedlungswellen nach Kuba gelangte. Beide Gründungspopulationen divergierten durch genetische Drift weiter, während sie Richtung Inselmitte wanderten, wo sie hybridisierten. Dort evolvierten sie zu einer Gruppe mit distinkten Charakteristika.

1. Introduction

The islands of the Caribbean are a hot spot of biodiversity. Their faunal composition comprises South, Central and North American faunas and island effects may favour the incidence of endemic species. Therefore, the Caribbean was placed third among 25 hotspots of biodiversity world-wide that were assessed for conservation priorities (Myers *et al.*, 2000; Conservation International, 2005).

High biodiversity applies also to the chiropteran fauna of the Caribbean. Here, bats are the most abundant and diverse mammals in terms of the number of species and the number of endemics (Griffith & Klingener, 1988; Hedges, 2002).

Studies on the biogeography of bats in the Caribbean have revealed diverse patterns of colonization and accompanying species diversification. Some species entered the Caribbean from South America and inhabited the Lesser Antilles (southern route), whereas others came from North America colonizing the Bahamas along the way (northern route). However, most seem to have entered the Caribbean from Central America (western route. (Baker & Genoways, 1978; Hedges, 2002). Dispersal of some species (e.g. *Natalus* spp.) occurred also from the Caribbean to the American mainland (Dávalos, 2004). Nonetheless, allopatric divergence, i.e. evolution in disjunctive geographic areas such as the islands of the Caribbean, is the major mode of speciation. Thus, subsequent diversification on the islands led to recent patterns of endemics and/or subspecies distribution.

One of the five most frequently encountered and widely distributed bat species in the Greater Antilles is *Macrotus waterhousii*, Gray, 1843 (Gray, 1843). It can be found on most of the islands of the Greater Antilles and it is likely that island populations of *M. waterhousii* represent distinct lineages, subspecies or even species. Thus, Caribbean *M. waterhousii* is believed to consist of a multispecies complex (Rehn, 1904; Koopman *et al.*, 1957; Anderson & Nelson, 1965; Buden, 1975; Griffith & Klingener, 1988; Murray *et al.*, 2008; Fleming *et al.*, 2009;

Murray *et al.*, 2009). I therefore investigated morphology, echolocation and foraging behaviour, social interactions and genetic diversity of Cuban *M. waterhousii* from different regions of Cuba.

I chose Cuba as a study area because it has at least two *M. waterhousii* morphotypes (Gundlach, 1872; Rehn, 1904; Silva Taboada, 1979) and because Cuba plays an important role in Caribbean biogeography. Cuba is a pivotal island for species diversity patterns in the Caribbean for two reasons. Firstly, it is the largest island of the Caribbean, covering 115 000 km², and is only 200 to 300 km from the mainland of America and thus to the source of colonization. Secondly, the proximity of neighbouring islands makes Cuba a hub for species dispersal within the Caribbean. To the east, Cuba's closest neighbour is Haiti at a distance of 94 km and to the south, Jamaica, at a distance of 157 km. Area size and proximity to colonization sources positively affect bat species diversity (Baker & Genoways, 1978; Ricklefs & Bermingham, 2008; Willig *et al.*, 2009). Indeed, Cuba has the richest bat fauna throughout the Antilles with the highest number of endemics (16), species (26 of 32), genus diversity (22) and guild diversity (6) (Baker & Genoways, 1978; Morgan, 2001). Apart from the endangered Cuban Solenodon (*Solenodon cubanus*) and the endemic species of Hutías (capromyid rodents), bats are the only indigenous terrestrial mammal. With the exception of blood-consuming vampires, which have become extinct (Morgan, 2001), the Cuban bat fauna comprises all known bat guilds, including narrow space gleaning bats, open space aerial foragers, edge space aerial foragers, and trawling bats. This diversity in foraging guilds is matched by the diversity in diets, which include insects, fish, fruit, nectar and pollen (Silva Taboada, 1979). Knowledge of Cuban bat diversity, including *M. waterhousii*, is almost exclusively based on extensive morphological investigations (Silva Taboada, 1979).

In current taxonomy (Simmons, 2005), *M. waterhousii* has one sibling species, *M. californicus*, which is found in northern Mexico and the south-western parts of the United States. *Macrotus waterhousii* is also widely distributed on the mainland, ranging from north-western Mexico to Guatemala and Belize.

Authors of previous studies described up to eleven morphotypes (for distribution ranges and names of geographical areas, see Figure 1) which were ascribed to various taxonomic levels (taxonomic classification of *Macrotus* Table 25 in the Appendix).

Figure 1: Distribution area of the genus Macrotus. Dark grey refers to M. californicus, black refers to M. waterhousii and light grey on the Mexican mainland refers to an area of overlap between the two species. Dotted lines indicate limits of distribution areas of morphotypes and white dots to the sample locality of reference individuals. Names in bold italic refer to morphotype name.

Macrotus belongs to the family of New-World, leaf-nosed bats (Phyllostomidae), the second most diverse bat family (Koopman, 1994). It preys mainly on large insects, but also consumes small vertebrates (pers. obs.) that it gleans directly from the ground or vegetation. *M. waterhousii* lives in mixed gender groups of various sizes and roosts in caves or manmade shelters where it seems to prefer ventilated areas (Anderson, 1969; Silva Taboada, 1979; Morgan, 2001; Avila-Flores & Meddelín, 2004; Genoways *et al.*, 2005).

In addition to a number of morphological studies, a few studies analyzing karyotypes, biochemical and gene-sequence data are available for this genus (Nelson-Rees *et al.*, 1968; Davis & Baker, 1974; Greenbaum & Baker, 1976;

Baker & Genoways, 1978; Straney *et al.*, 1979; Baker & Bickham, 1980; Van Den Bussche, 1992; Freeman, 2000; Murray *et al.*, 2008; Fleming & Murray, 2009; Fleming *et al.*, 2009; Muscarella *et al.*, 2011). Recently, two studies investigated the molecular diversity of *M. waterhousii* across the Caribbean islands (Fleming *et al.*, 2009; Muscarella *et al.*, 2011). However, both studies had small sample sizes for the Cuban *M. waterhousii* population. Nevertheless, these studies strengthened the notion of a multispecies complex for *M. waterhousii* in the Caribbean island including Cuba where variability in morphology has been studied. The Cuban subspecies, *M. waterhousii minor* (Gundlach, 1864), was reported to be the smallest of all the morphotypes including a very small morphotype occurring in the western part of the island and a larger morphotype in the eastern part. These morphotypes were distinguished by size differences in forearm length and skull and dental measurements, as well as by differences in fur coloration and strength of claws. Generally the small western morphotype was more slender in appearance of (Gundlach, 1872; Rehn, 1904; Silva Taboada, 1979). The geographical separation of these two morphs is distinct and has been described to occur in "a step-like manner" (Silva Taboada, 1979; see white line dividing Cuba in Figure 1) with no intermediate forms in the central region of the island. A hypothesis concerning the historical colonization routes of *M. waterhousii* was put forward by some researchers to explain the morphotype pattern across the Caribbean in general (Koopman *et al.*, 1957), but also to explain the Cuban pattern (Anderson & Nelson, 1965; Buden, 1975; Koopman, 1989). According to this hypothesis, *M. waterhousii* entered the Caribbean from the mainland of Mexico (north-western route) in two major invasions (Buden, 1975; Baker & Genoways, 1978; Griffith & Klingener, 1988). The first invasion brought a population of smaller specimens from the Yucantán Peninsula, indicated as *bocourtianus*-morph in Figure 1, directly to the western part of Cuba. A second invasion stock consisted of the more robust *mexicanus*-morphotype that entered the Caribbean via Jamaica and Haiti before inhabiting eastern Cuba. Colonisation took place during the late Pleistocene which was dominated by multiple sea-lowering events which facilitated dispersal to Cuba or Jamaica (Koopman *et al.*, 1957; Baker & Genoways, 1978; Itturalde-Vinent & Macphee, 1999). If two divergent populations meet again as suggested for the Cuban *M. waterhousii*, several scenarios are conceivable. If the divergence between the two forms was pronounced before they arrived on Cuba they could have simply remained apart as two morphotypes. Conversely, if the process of divergence had not proceeded very far, the populations may simply have interbred and merged with one another. Explanations for current variability might then be found in environmental conditions such as average temperature or rainfall which change gradually with latitude and/or longitude. Some aspects of the organism, e.g. size, may then also vary gradually and continuously across the area, a regular change known as a cline (Cox & Moore, 2000). If barriers to hybridization have arisen, interbreeding might be restricted. However, those barriers can take many forms.

Cuba has no geographic barriers promoting population, i.e. morphotype, separation. In addition, the reported lack of hybridisation indicates an isolating barrier of some kind. Such barriers can develop on the basis of geography, as well as ecology, ethology or genetic drift and these factors may act in concert to produce divergence. Some of which might be cryptic. Bats are striking examples of how these manifold forces can produce cryptic speciation. Morphologically similar species were found to be genetically distinct, i.e. cryptic species were discovered as genotypes (Mayer & Von Helversen, 2001; Kiefer *et al.*, 2002). Another species-elucidating approach was the analysis of echolocation calls; morphologically similar species differed remarkably in their call design, revealing phonotypes that correlated with genetic differences (Jones, 1997b; Russo & Jones, 2000; Mayer & Von Helversen, 2001; Kiefer *et al.*, 2002; Jones & Barlow, 2004). This clearly demonstrates the need for broad approaches for adequate species recognition.

I aimed at providing a detailed assessment of Cuban *M. waterhousii* population variability. I therefore chose a comprehensive approach that included morphological, acoustic, and genetic analyses. I addressed the following questions: a) whether described Cuban morphotypes are a result of an ecological cline or a two-invasion scenario, and b) whether variability developed under positive selection or genetic drift by testing the biological relevance (i.e. adaptive significance) of differences between the two morphotypes.

The first aspect of my study was a revision of morphological differences in *M. waterhousii* based on exemplars from a wide geographical range to verify the existence of phenotypic differences and the underlying reason for such differences. To determine whether differences in morphology affect flight and foraging behaviour, I examined the manoeuvrability. A main part of this thesis was the analysis of echolocation behaviour. I compared the acoustic structure of echolocation pulses emitted in a standardized situation and tested performance in experiments. As ethological barriers have the potential to separate populations, I studied the social interactions of bats from different geographical areas in captivity. The genetic variability was analyzed to reveal a genetic pattern across Cuba which may match, on a molecular level, the phenotypic pattern.

Morphology has been the predominant tool for species recognition in the past, but as mentioned above, the identification of cryptic species shows that morphology is not always a reliable tool for taxonomy. The morphology of bats is adapted to the respective habitat and foraging strategy. This is obvious in wing shape and related aerodynamic features that influence flight speed and manoeuvrability (Norberg & Rayner, 1987), but also in other morphometrical parameters like ear characteristics or dentition. As the ecology of the habitat can shape morphology, similarities in morphology may reflect guild membership rather than taxonomic relatedness. At the same time, morphological differences can restrict the access of a species to matching niches. The morphology of *M. waterhousii* is adapted to its foraging strategy in terms of prominent large ears and wing morphology, which enable the species to manoeuvrable flight, hovering and vertical uplift from the ground.

Hypotheses concerning morphology:
1a. Corresponding to previous findings, morphological variability results in an eastern versus a western morphotype.
1b. Morphological differences affect the manoeuvrability of morphotypes.

Ecology also shapes echolocation of bats. Most bats that feed predominantly on insects rely on echolocation for orientation and foraging. Bats can be classified into guilds by their echolocation pulse design according to their preferred hunting habitat and foraging strategy (Schnitzler & Kalko, 2001; Schnitzler *et al.*, 2003; Surlykke *et al.*, 2003; Von Helversen & Von Helversen, 2003; Boonman & Ostwald, 2007). *Macrotus waterhousii* is a gleaning bat species orientating, hunting, detecting, locating and capturing prey within vegetation and is able to pick up prey directly from the surface. This hunting situation is acoustically challenging because the distance from the bat to its prey is short. Thus, prey can be hard to discriminate within the cluttered background and echo and pulse can overlap, i.e. the prey echo can be masked by the vegetation echo. The short duration of echolocation of gleaners cope with the masking problem. A low intensity, a multiharmonic structure covering a broad frequency band and reaching high frequencies increases the resolution of structures within the clutter. Echolocation pulses of *M. waterhousii* are typical for phyllostomids and gleaners. They are short in duration, cover a broad frequency range and are high in frequency. Like many gleaners, *M. waterhousii* may additionally use other sensory cues for hunting. It has been shown that *M. californicus* uses vision when possible and exploits prey generated rustling noises (Bell, 1985; Bell & Fenton, 1986). Although echolocation structure has similar characteristics among guild members, it is often species specific (Fenton & Bell, 1981; Barclay, 1983; Ahlén, 1990; O'farrell & Gannon, 1999; Russo & Jones, 2002; Macías *et al.*, 2006) and can serve for species identification.

Besides some exceptions (Jacobs *et al.*, 2007), frequencies of echolocation pulses can be correlated with morphological parameters, in particular size

(Novick, 1977; Heller & Helversen, 1989; Barclay & Brigham, 1991; Jones, 1996) as the length and strength of vibrating parts and connected muscles of the vocal tract determines the oscillation characteristics and therefore the frequency of sounds. This allometry has been shown within several bat families and indicates the evolution of morphology and echolocation as a complex.

Hypotheses concerning acoustic orientation and foraging:

2a Echolocation pulse characteristics differ between individuals from geographic areas and correspond to morphotypes.

2b *M. waterhousii* uses predominantly passive listening to prey generated sounds to detect prey. If so, bats should prefer feeding dishes with most conspicuous rustling noises of mealworms in a three-choice experiment.

2c. Acoustic differences in echolocation pulses of bats from the west or the east affect prey capture success under different clutter demands.

Another approach when investigating diverging populations is to analyze social behaviour. Like many other bat species *M. waterhousii* lives in social communities with conspecifics. Group-living requires communication between individuals to govern interactions. A prominent sensory channel used for communication by bats is the acoustic channel. Rich repertoires of social calls are known from several bat species (Leippert, 1994; Behr & Von Helversen, 2004; Kulzer, 2005; Ma *et al.*, 2006). Usually, social calls are context specific (Barclay *et al.*, 1979; Leippert, 1994; Boughman & Wilkinson, 1998; Pfalzer & Kusch, 2003; Behr & Von Helversen, 2004; Davidson & Wilkinson, 2004; Schmidt, 2005), they can convey individual signatures (Gelfand & Mccracken, 1986; Balcombe & Mccracken, 1992; Leippert *et al.*, 2000), group affiliation (Esser & Schmidt, 1989; Russo & Jones, 1999; Doerrie *et al.*, 2001), group-specific and species specificity (Barclay *et al.*, 1979; Lundberg & Gerell, 1986; Scherer & Wilkinson, 1993; Boughman & Wilkinson, 1998; Pfalzer & Kusch, 2003). In terms of variability and speciation, it is essential to maintain efficient communication and to recognize conspecifics correctly. Thus, there is a high selection pressure on intraspecific communication (Jones, 1997a; Russo & Jones, 2000; Kingston *et al.*, 2001), in particular, when populations are in the process of divergence or species face a second contact (Grant & Grant, 2002). So far, the social structure and social behaviour of *Macrotus* are barely known and information is scattered throughout the literature (Dobson, 1878; Gould *et al.*, 1973; Gould, 1975; Silva Taboada, 1979; Leblanc, 2001). Berry and Brown (1995) reported that males of *M. californicus* show a specific behaviour during the mating season. Each male occupies a small area (a lek) and displays wing flapping with accompanying chirp calls. Groups of both *Macrotus* species were found to consist of both sexes and seasonally of one sex (Anderson, 1969; Silva Taboada, 1979; Bell *et al.*, 1986; Berry & Brown, 1995), and it is known that *Macrotus* regularly visits the same night roosts where it feeds on its prey (Anderson, 1969; Silva Taboada, 1979; Bell *et al.*, 1986) and may interact with conspecifics. If populations of *Macrotus* had undergone differentiations of social behaviour in the form of acoustic dialects or in behavioural displays, individuals from different localities might interact less frequently, or differently, when compared to interactions between locality members.

Hypotheses concerning social interactions:
If the divergence between the two morphotypes is maintained through an ethological barrier then
3a. individuals have a stronger affiliation to locality mates. Individuals caught in the same geographic area prefer perches in proximity to each other in the flight room.
3b. individuals caught in the same geographic area allogroom more frequently than individuals from different geographic areas.
3c. social calls emitted in the same context, but from individuals from different geographic areas, differ in their acoustic structure.

With the advances in molecular methods, phylogenetic analyses elucidated, renewed, corroborated and expanded the knowledge of chiropteran taxonomy essentially (Simmons, 2005). Molecular methods can be applied to all levels of taxonomy and the high number of newly identified species is certainly largely the result of phylogenetic analyses. An analysis of the genetic divergence within *M. waterhousii* across Cuba provides another powerful tool with which to assess the variability of this species and to correlate the results of morphology, echolocation and social behaviour to it. As my intention was to contribute a sample set of comparable sequences of Cuban *M. waterhousii*, I decided to sequence fragments of the mitochondrial Cytochrome-b (cyt-b) gene and a non-coding control region (d-loop) of the mitochondrion as these markers had already been studied in *Macrotus* before (Hoffmann *et al.*, 2008; Fleming *et al.*, 2009). The combination of both markers may help to determine whether the variability of Cuban *M. waterhousii* is a product of recent processes or has existed for a longer time as these markers differ substantially in their mutation rate (Howell *et al.*, 1996) with the d-loop evolving more quickly than the conservative cyt-b.

Besides the mode of speciation at the beginning of species divergence, i.e. natural selection or genetic drift, progressing speciation processes can create variability between populations on a continuous landmass reflecting either specific adaptations (natural selection) or effects of isolation (genetic drift). Therefore, I applied a divergence-by-geographic distance correlation to reveal the effect of longitude, i.e. isolation, on *M. waterhousii* variability.

Hypotheses concerning genetic variability:
4. The pattern of genetic variability matches variability of morphology and echolocation. The structure of haplotypes and the topology of phylogenetic trees resulting from cyt-b and d-loop markers are congruent. D-loop analysis results in a higher resolved structure.

2. Material and Methods

2.1 Data base

Field trips were done in May, June, July and September in 2008, in December 2009, January, February and March 2010. All caves were visited during daytime and sampled areas ranged from the most western point to Santiago de Cuba covering most of Cuba's extension (Figure 2).

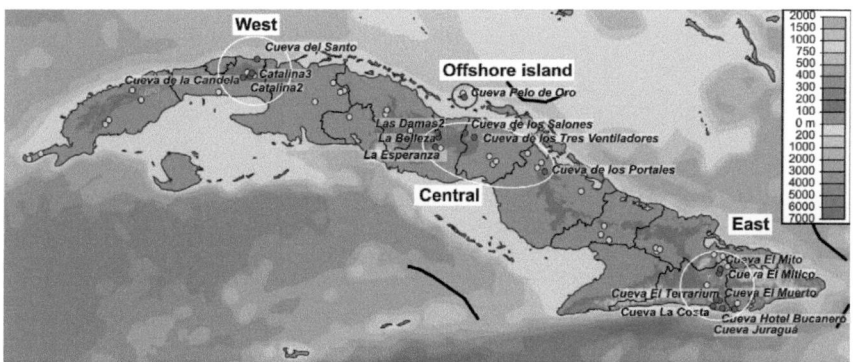

Figure 2: Location of caves sampled in this study. Dots indicate all sampled caves in 2008 and 2009/2010. Grey dots refer to caves where Macrotus was absent or where I could not capture it (the two westernmost caves). Red dots indicate caves where specimens of Macrotus were sampled and names of these caves are given. The four major sample sites, the west, central, offshore island and the east, are circled. Coloration of the map indicates topographic heights in meters above (green and brown) and below (blue) the sea level as given in the legend.

The data presented here are based on individuals from fifteen caves representing four geographical areas, namely the west, the central, the eastern area and an offshore island north-central to Cuba (Tables 26 and 27 in the Appendix give distances between sample sites and caves).

Bats were captured inside the cave and at the entrance by hand nets (butterfly nets, 50 cm in diameter, 1 mm mesh, 80 cm depth by Bioform, Nuremberg, Germany) and/or mist nets (75 denier/2 ply, 38 mm mesh, 2.6 m x 6 m and/or 2.6 m, 4 shelves by Avinet, Inc., Dryden NY, USA). Prior to recordings of echolocation pulses and measuring morphological characters, bats were kept in groups of up to five inside soft cotton bags (24 cm x 18 cm). Most of the captured bats were released immediately after data collection. I marked released animals by shaving an edge into the fur to recognize them in case of recapture. Female bats were immediately released if they showed signs of pregnancy or if they were lactating. To obtain ecological data of males which might be relevant for mating I determined the development of glandulae which are positioned right behind the ears of male *M. waterhousii*. Complete ossification of the finger joints confirmed that bats were adult. Small tissue samples for DNA analyses were obtained from biopsy punches (3-4 mm²) of the uropatagium (interfemoral membrane).

Some individuals of western, central and eastern sample site were brought to an animal facility at the University in Havana. I kept a first group caught in the west and in the east from May to November in 2008 and another group caught in the west, in the central area and in the east from January to July 2010. Before releasing the bats into the flight room of the facility (for a detailed description of the facility see section 2.3.1) they were kept in transportation cages to habituate them to the novel food source (larvae of *Zopohobas morio*, *Tenebrio molitor*, *Galleria mellonella* and *Periplaneta americana*) by feeding them by hand. The permanently provided water bowl was quickly accepted by all bats. One night before releasing the bats into the flight room I marked them individually by a collar with a letter or a number.

2.2 General methods

2.2.1 Audio recordings and analyses

Echolocation calls and social calls were recorded by a Bat detector (D 1000 X, Pettersson Elektroniks, Uppsala, Sweden) and automatically saved on internal data storage (CF cards) in .wav format. Recordings had a sampling rate of

384.00 kHz and a 16-bit depth. For acoustical analysis of calls, I used two software programs: BatSound Pro 3.31 (Pettersson Elektronik AB, Uppsala, Sweden) and AviSoft SASLab Pro 4.51 (Avisoft Bioacoustics, Berlin, Germany). For playing back a typical echolocation pulse to the clutter backgrounds in the respective experiment (see section 2.3.2.4), I used an ultrasound loudspeaker (10 - 110 kHz output frequency range, max. output 100 dB at 1 m distance. Type L140, Pettersson Elektroniks, Uppsala, Sweden).

A set of 16 acoustic measurements containing time, intensity, and frequency parameters was chosen to characterize echolocation pulse structure. Calls were characterized by their number of harmonics, their duration, their peak- start- and end frequency, their bandwidth, prominent harmonic and modulation rate. For definitions, program settings and method of measuring see Table 1. In multiharmonic echolocation pulses of *M. waterhousii*, second and third harmonics are often the most pronounced harmonics and they overlap partially in frequency. Reasons for changes in peak frequency within one individual can therefore simply be a result of shifting the main energy from the second to the third harmonic, or vice versa. To exclude this effect, I decided to isolate the second harmonic and analyze it separately as it was in 87.2% of all 687 echolocation calls the prominent harmonic (third harmonic and fundamental were sometimes also prominent. See emission of the fundamental in echolocation pulses Table 28 in the Appendix).

Table 1: List of acoustic parameters measured in echolocation pulses. Number and name of parameter is followed by a description and method of acquisition. [1] indicates those parameters included in the multiparametric analyses. Table continued on next pages.

Parameter	Description	Method
1) Signal-to-noise ratio (dB)	Mean level of pulse amplitude minus mean level of background noise amplitude (dB). Measured at 2 ms distance to pulse, max. 15 ms before pulse onset and max. 15 ms after pulse end. Only pulses with signal to noise ratio between 11 dB and 40 dB were analyzed.	Oscillogram and power spectrum, Hanning window, 128 samples.

Table 1 continued.

2) Number of harmonics	Number of harmonics of each pulse.	Sonagram, high-pass filtered, Hanning window, 128 samples.
3) DurP (ms) [1]	Pulse duration. From onset of the pulse to end of the pulse.	Oscillogram, high-pass filtered.
4) PeakfP (kHz) [1]	Peak frequency entire pulse. Frequency of maximum energy in the pulse.	Power spectrum, high-pass filtered, Hanning window, 128 samples.
5) Prominent harmonic	Harmonic(s) which differ <6 dB from peak frequency.	Power spectrum, high-pass filtered, Hanning window, 128 samples.
6) StartfP (kHz)	Start frequency. At -25 dB above the peak frequency on highest harmonic.	Power spectrum, high-pass filtered, Hanning window, 128 samples.
7) TermfP (kHz) [1]	Terminal frequency. At -25 dB below the peak frequency on lowest harmonic.	Power spectrum, high-pass filtered, Hanning window, 128 samples.
8) eff.BandwP (kHz) [1]	Effective bandwidth of entire pulse. Measured from the peak frequency of the lowest harmonic to the peak frequency of the highest harmonic.	Derived value.
9) BandwP (kHz)	Bandwidth of entire pulse. Margin between start and terminal frequency.	Derived value.
10) ModRate (kHz/ms)	Modulation rate. Bandwidth divided by pulse duration.	Derived value.
11) Frequency-time-course 2nd Harmonic	Time and frequency values of the second harmonic were extracted.	Spectrogram, high-pass filtered, Hanning window, 128 samples.
12) 2nd HarmPeakf (kHz) [1]	2nd harmonic peak frequency. Frequency of highest energy of the second harmonic.	Power spectrum, high-pass filtered, Hanning window, 128 samples.
13) 2nd HarmStartf (kHz)	2nd harmonic start frequency. Highest frequency at -25 dB above peak frequency of the second harmonic.	Power spectrum, high-pass filtered, Hanning window, 128 samples.
14) 2nd HarmTermf (kHz) [1]	2nd harmonic terminal frequency. Lowest frequency at -25 dB below peak frequency of the second harmonic.	Power spectrum, high-pass filtered, Hanning window, 128 samples.

Table 1 continued.

15) **2nd HarmBandw (kHz)**	2nd harmonic bandwidth. Difference between start and terminal frequency of the second harmonic. Extracted values from the frequency-time-course.	Derived value.
16) **2nd HarmModRate (kHz/ms)**	2nd harmonic modulation rate. Bandwidth of the second harmonic divided by pulse duration.	Derived value.

Due to very short pulse durations (as a function of sound source distance to the microphone) which made it difficult to achieve a sufficient frequency resolution, I did a zero padding by adding 6 s of silence around the pulse. With a fast Fourier transformation rate of 128 in a Hanning window and a temporal overlap of 99% I achieved an increased frequency resolution of around 1953 Hz (and 2930 Hz bandwidth). Only recordings with a sufficient signal to noise ratio (>11 dB) were analyzed and, in addition, only adult animals with complete morphological data sets entered the analysis.

2.2.2 Video recordings and analyses

Experiments were videotaped by a camcorder (Sony DCR-HC 85E) with an activated nightshot function), stored on miniDV cassettes (Panasonic, SP modus, 60 min) and digitised on a laptop (Pinnacle studio version 10.8.0.4641, high DV-recording, 25 frames/s, sample rate 48 kHz. AVI format: length and width 720 x 576 pixel or MPEG format: length and width 720 x 576 pixel). Video analysis of experiments was carried out by using the software InterAct (Version 8.0.4, Mangold International GmbH, Germany). Only scenes with zero lost frames were analyzed.

2.3 Experiment specific methods

2.3.1 Field studies

2.3.1.1 Morphometric measurements

To minimize stress effects on acoustic behaviour, measurements of morphological parameters were done after the call recordings. In addition to a set of 18 body measurements, see Table 2 (spring balance: 30 g capacity with 0.25 g division by Pesola, Switzerland. Sliding calliper: 0.01 mm precision by Hogetex/Kometex B.V., Netherlands) sex, age, reproductive status, and fur coloration was noted. For wing measurements in the field, I took photos (Nikon D40, 2048x1536 pixels) of bat wings which were later analyzed on a laptop.

Table 2: List of morphometric parameters taken for each individual. Number and name of parameter is followed by a description and method of acquisition. Measurements of parameters were taken from the right body side. Parameters thumb length and tail length were acquired in 2009 but not in 2010. [1] *Parameters included in multiparametrical analyses. Parameters 1) - 5) were obtained by inspecting individuals by eye, parameter 6) by a spring balance, parameters 7) - 15) by a sliding caliper and 16) - 18) by measuring on a laptop from a photo. Table continued on next page.*

Parameter	Description
1) **Sex**	Male or female. Inspection of outer sexual characteristics.
2) **Age**	Adult or juvenile. Complete ossification of finger joints in adult individuals.
3) **Reprod. status** ♀	Inspection of belly size and secretion activity of mammary glands. Categories were: Pregnant, last stage of pregnancy, lactating, post lactating, none of these. See Photo of a lactating female (Figure 25 in the Appendix).
4) **Reprod. status** ♂	Wet head glandulae with fur; wet glandulae and bold-headed, no wet glandulae but bold-headed, no glandulae visible. See photo of a bold-headed male (Figure 26 in the Appendix).
5) **Fur coloration**	Brownish, grayish, reddish.
6) **Weight (g)**	Body weight of the individual. Bat weighed in a cotton bag. Weight of cotton bag subtracted.

Table 2 continued.

7)	**Forearm length (mm)** [1]	Measured from the end of the ulna to the wrist.
8)	**Ear length (mm)** [1]	Measured from basal outer sharp bend of the ear to the tip of the ear.
9)	**Tragus length (mm)** [1]	Measured from the outer basis of tragus to the tip.
10)	**Tragus width (mm)** [1]	Measured at the basal widest point of the tragus.
11)	**Nose leaf length (mm)** [1]	Measured from the basis of the nose leaf to the tip.
12)	**Nostrils distance (mm)** [1]	Measured from center of one nostril to center of second nostril.
13)	**Snout width (mm)** [1]	Measured at outer basis of lower canini.
14)	**Thumb length (mm)**	Measured at basis of first phalanx to basis of nail.
15)	**Tail length (mm)**	Measured from the basis of the tail to the tip.
16)	**Wing span (mm)**	From tip of left wing to tip of right wing. Photos of extended right wing and half of the bats body on scale paper. Bats were laid on their back, legs were straight in the longitudinal axis. Photos were taken in 25 cm distance in 45° angle. Measured value was multiplied by two.
17)	**Wing area (mm²)**	Measured from the contour of the right wing. Measured value was multiplied by two.
18)	**Aspect ratio**	Square of wingspan divided by wing area.

2.3.1.2 Orientation pulses in a standardized situation

I recorded echolocation pulses in an orientation situation under standardized conditions in the field. The bat was positioned in 1 m distance to a sound absorbing screen of foam material (130 cm x 200 cm x 1 cm) to reduce effects of different backgrounds on echolocation call structure. A microphone in the centre of the screen was connected to the bat detector via an extension cable and the bat was held 1 m from the ground and 50 cm in front of the microphone (inserted through a small hole in the foam). To reduce the freedom of movement, each bat was placed between two pieces of soft foam with a body-shaped recess in the lower piece. This was necessary in order to minimize direction-dependent changes in the spectral structure of echolocation calls. I recorded spontaneously emitted echolocation calls for 5 seconds. If the bat was not spontaneously emitting echolocation, I stimulated sound emission by slightly moving the bat. In this case, recordings started after 10 seconds.

An individual was included into the acoustic analysis: if it was adult, if I obtained both, morphological data and echolocation recordings, and if at least five pulses per individual were analyzable. For multivariate analyses, I excluded outliers and extremes.

2.3.2 Behavioural studies in the lab

2.3.2.1 Laboratory facilities: Flight room and experimental room with experimental cage

I kept mixed-sexed groups of adult individuals from three geographical sites. Group size varied from 7 to 23 individuals. Bats were kept from May until November 2008 and from January to July 2010 in a flight room (for a list of bats and their arrival dates see List of individuals kept in the lab Table 29 in the Appendix). This room was a separate part of a large facility (7.50 m in length x 3.65 m in width x 2.90 m in height. For a sketch of the facility, see Figure 3). I changed the light regime (darkness from 09:00h to 22:00h) to conduct the experiments and to record social interactions within the activity phase of the bats. Four red-light bulbs (15W) gave light continuously. One white light bulb (15W) illuminated the flight room during artificial daytime. A small window guaranteed influx of fresh air and an occasionally used fan brought additional circulation of air. In one corner, I constructed a removable roost-site to give bats the possibility to withdraw. Water access was ad libitum, whereas food was provided exclusively in the experimental cage after a habituation phase. Regular food was larvae of *Zopohobas morio*, *Tenebrio molitor*, *Galleria mellonella* and *Periplaneta americana*. In addition, newborn mice were fed at least twice a month.

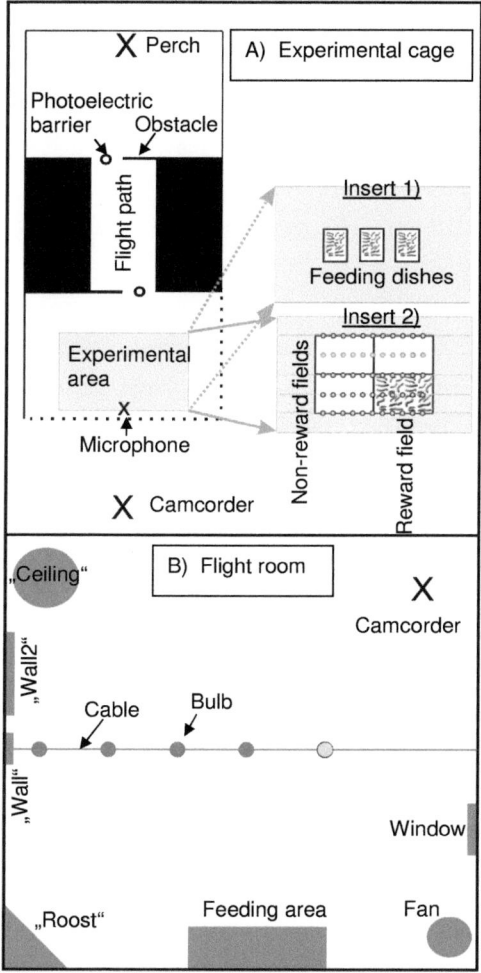

Figure 3: Sketch of the facility. Upper part shows the part of the facility in which the experimental cage was set up. Shown are the three parts of the experimental cage and indicated are setups for the experiment on manoeuvrability (in the central part), for the experiment on passive listening (in the frontal part. insert 1) and for the experiment on prey capture success in clutter (in the frontal part. Insert 2). The part below gives an overview of the flight room where bats were permanently housed. Indicated are main areas in the flight room (also referring to preferred areas as analysed in the part about spatial allocation).

The experimental cage was 3.00 m in length x 1.50 m in width x 1.00 m in height and set up in the facility next to the bats´ flight room. It was made of black nylon fabric with a smooth surface lest the bats could not attach on any other place than a fixed perch in the rear part. The cage was subdivided into three equally sized parts (1.00 m each); the rear part contained a perch made of wire mesh (17 cm x 19 cm). This part could be curtained-off by a nylon fabric. The central part was used to set up the obstacle course in the experiment on manoeuvrability and the frontal part was used for experiments on passive listening and prey capture success under different clutter conditions. Front and sides of the frontal part were made of transparent plastic tarp to allow video recordings from outside. The camera was positioned at a fixed distance of 1.00 m to the ground and 1.50 m to the front of the flight cage. To achieve a sufficient videos quality, I illuminated the cage with infrared spots pointing to the perch in the rear part and to a focused spot of the respective experiment. For accompanying audio recordings, the microphone of the bat detector was put on the ground in the frontal part and directed in a 70° angle to approaching bats at about 30 cm above the ground to record echolocation calls shortly before prey capturing. To reduce echoes of emitted calls, I covered the ground of this part with a soft foam mat (1.50 m x 1.00 m x 0.05 m). The experimental room was darkened during experiments (+/- 0.001 lux) and only illuminated between the trials by two red light bulbs (15 W) at both sides of the experimental cage.

2.3.2.2 Training of the bats

To conduct experiments inside the experimental cage, bats had to get habituated to the cage and to be trained for experimental tasks. The training procedure consisted of three stages; in the first stage, each bat was placed manually on the perch in the rear part which was separated by closing the curtain. Bats had at maximum of 20 min to explore this part and to re-discover and memorize the perch. If a bat could not find the perch and began to land on the ground, it was manually placed to the perch again. This was repeated until the bat had no difficulties anymore to find the perch. In the second stage, the curtain was lifted and bats could explore the entire flight cage. If a bat found an alternative perch, for instance a seam, or landed on the ground, it was chased-off by gently clapping hands. Again, this was repeated until the bat returned without errors to its perch after exploration flights. In the third stage, a feeding dish full of insect larvae was presented in the experimental area in the frontal part. The ground of the dish was covered by paper to make it easier for the bats to discover this food source as moving larvae produced loud rustling noises on this surface detectable by bats (see following section). If a bat could not make it to this stage because it stopped flying or did not approach or land in the feeding dish, a second bat which already achieved some routine was put as a tutor into the cage to encourage the other bat. After a period of two up to eight weeks all individuals had accomplished the training procedure.

2.3.2.3 Passive listening

This experiment was done in 2008 with five bats from two different capture sites (individuals "2", "5", "6", "R" and "S"). This experiment on prey acquisitions strategies was conducted to see to which extent the Cuban *M. waterhousii* exploits prey generated sounds and if it uses echolocation when acoustic cues through prey generated sound are available.

I defined a central area of 90 cm x 60 cm on the floor of the frontal part as the "experimental area" where the experiment was set up. On the video, the perch (starting point of the bats), the central part (flight path of the bats), and about 5/6 of the focal experimental area without the upper sixth of this part were visible. Three metal feeding dishes were prepared for this experiment (18 cm x 25 cm each): the ground of one was covered with a thin plastic foil, the second was covered with paper, and the third was left blank. *Tenebrio* larvae creeping on these surfaces caused different rustling noise profiles of the dishes. To characterize rustling profiles I recorded ten living *Zophobas* larvae in each of the dishes in a distance of 30 cm and did one recording without larvae to compensate for the frequency response of the recording system. I analyzed the power spectrum of a 3.5-second interval to measure creeping-sound-bandwidth above the background noise level and calculated means of relative sound levels above the background noise level. Dishes differed gradually in the intensity of rustling noises caused by creeping of larvae (see Figure 4) as their inlay was plastic (-58 dB "most conspicuous dish"), paper (-59 dB "mid-level conspicuous dish") or metal (-61 dB "most inconspicuous dish").

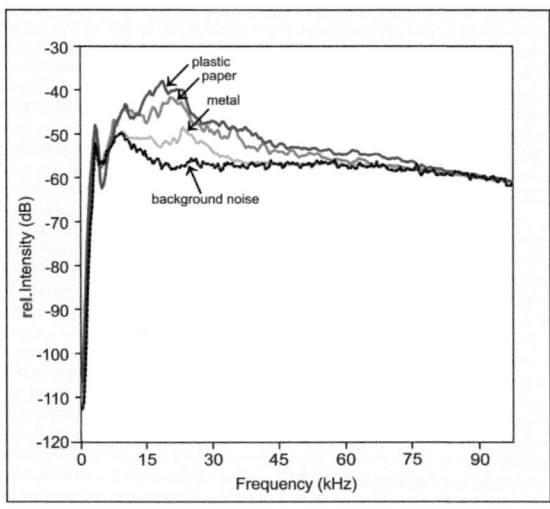

Figure 4: Rustling noise profiles of the three presented feeding dishes in comparison to the background noise. Rustling noises generated by creeping Tenebrio larvae were recorded with 30 cm distance. Different surfaces of feeding dishes caused noise differing in frequency and intensity. The black line represents background noise level. The light grey line represents rustling of creeping larvae on metal, the medium grey line rustling of larvae creeping on paper and the dark grey line rustling level of larvae creeping on plastic.

The bats could not make use of olfactory cues to discriminate between the three dishes as the amount and type of prey was comparable in all presented conditions. I started experiments during the darkness phase of the bats (09:00 h to max. 22:00 h). All three feeding dishes were put next to each other, separated by 10 cm in the experimental area in the frontal part while the bat was at its perch in the curtained-off back part (see Insert 1 in Figure 3). Ten *Zophobas* larvae were put in each dish. The experiment began by lifting the curtain. After ten minutes and if bats were at the perch the curtain was closed again. The order of feeding dishes was changed randomly. Again, ten Zophobas larvae were put into each dish. Each bat did three trials per day and I analyzed nine trials of each bat. From the video sequences, approaches, landings and prey captures for each dish were analyzed (see following table).

Table 3: Parameters used for the analysis of the experiment on passive listening.

Parameter	Definition
Approach	Approach from the perch to the feeding dishes in the frontal part.
First approach	First approach of a bat from the perch to the feeding dishes.
Landing	Landing in one of the dishes.
Landing after first approach	First approach with subsequent first landing on a dish (yes or no).
Prey capture	Bat successfully captured prey after landing in a dish.
Incorrect landing	Bat landed somewhere in the experimental area, but not in a dish.

To avoid pseudo replication, the events "approach to dish", "approach with or without landing" and "first approach to dish within a trial" were analysed separately. I controlled for position preferences and checked improved performance over time (Cochrans-Q test). To analyse material preferences and position preferences Chi^2 tests of frequencies were calculated. Basis of calculations and analyses were individuals. I did audio recordings to ascertain that bats used echolocation during prey capture events and to study whether they differ acoustically from echolocation calls in orientation situations.

2.3.2.4 Prey capture success under two different clutter conditions

Fourteen individuals were trained to find silent and motionless prey (dead *Tenebrio* larvae) in the absence of visual cues on the ground in the experimental cage.

The experiment was set up again in the central area of 90 cm x 60 cm on the floor of the frontal part in the "experimental area". Dead mealworms were presented within three different clutter types; a moderate, a high clutter condition and a condition with no clutter background. I put five stripes of artificial flower garlands with a fixed distance, but maintaining similar distances between both clutter types, to each other on the floor. One flower garland consisted of 2.3 cm x 2.3 cm x 2.2 cm flowers, representing the moderate clutter condition. The other flower garland consisted of 6.8 cm x 5.2 cm x 6.0 cm flowers, representing the high clutter condition.

To evaluate the acoustical differences in the echoes perceived by the bats in the presence of different clutter background, I played back an average echolocation pulse and recorded the echo. It was played back in 30 cm distance at a 70° angle (a regularly observed hovering position) directed on the ground with or without respective flower garlands. Time between pulse emission and receiving the echo was different between the three clutter conditions (Figure 5). No clutter: 2.57 ms, moderate clutter: 2.42 ms and high clutter: 2.10 ms, measured on the on highest point of the flowers. Accordingly, the covered distance of the acoustic signal differed. No clutter: 44.3 cm, moderate clutter: 41.8 cm, and high clutter: 36.2 cm.

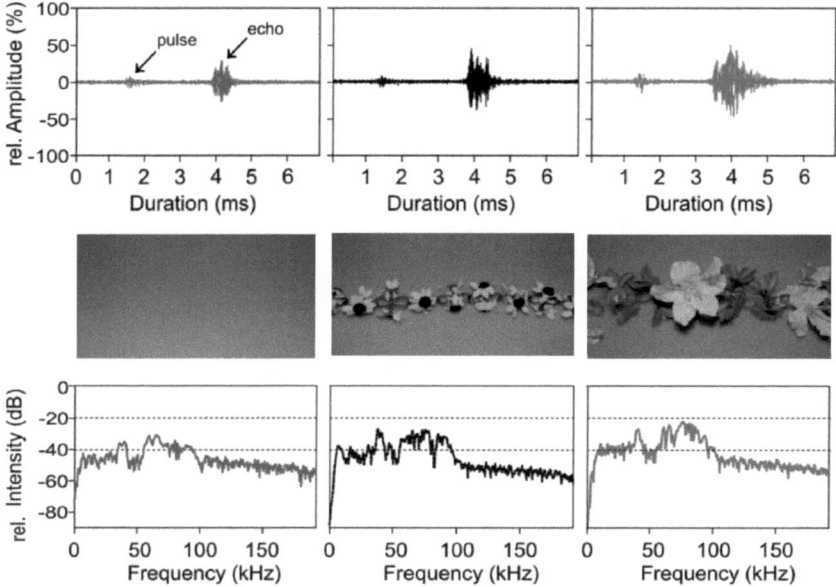

Figure 5: Echo characteristics of the three clutter types: no flowers = no clutter, small flowers = moderate clutter condition, large flowers = high clutter condition. The upper part shows oscillograms of playback recordings including the played back pulse, echo onset and echo duration. The central part shows pictures of the three clutter types and the lower part shows power spectra of the three clutter types. Dotted lines mark an area of -40 to -20 dB to indicate spectral intensity differences between the clutter types. Note that echoes differ up to 20 dB in spectral intensity.

The experimental area was subdivided into four equally sized fields (45 cm x 30 cm) to ensure that all experimental individuals with different wing spans had comparable conditions. A fixed amount of 70 dead mealworms were put scattered in one of these fields for each trial. The four test fields were noted as "reward field" or "non-reward field" (see Insert 2 in Figure 3). The video perspective again included the perch (starting point of the bats), the central part (flight path of the bats), and about 5/6 of the focal experimental area. Each bat did one session per day and one session consisted of three trials (plus initial feeding dish trial). One trial lasted five minutes and each bat did all four trials consecutively. In the first trial, I presented the feeding dish in one, randomly chosen, test field, as bats had to be motivated.

I analysed 6 sessions of each clutter condition per bat each consisting of 3 trials. Although bats had a maximum of five minutes per trial, I noted events from the video sequences until the bats had landed 15 times. Recorded events were: number of approaches to each field, number of landings in each field, number of hovering events prior landing and capture success. As there was only one reward field but three non-reward fields, expected frequencies for Chi² test were adapted.

After successful training of which I analyzed the last three days as the non-clutter condition and a three-day break of experiments bats were confronted with the clutter conditions. I started with the moderate clutter condition and analyzed the first three sessions (inexperienced) and, after a three-day break, the last three sessions (experienced). After, again, a three-day break, on the seventh testing day, bats had to find prey in the high clutter condition (inexperienced) on three subsequent testing days. After, again, three days break the last three sessions per bat (experienced) were analyzed.

2.3.2.5 Manoeuvrability

Bats had to fly through an obstacle course in the central part of the experimental cage (see Figure 3). First, bats were trained to fly from their perch through a 50 cm wide and straight corridor in the central part to the feeding dish in the frontal part and back. The video perspective included the entrance to the obstacle course, the corridor, the exit of the obstacle course, the frontal feeding area and the upper sixth of this frontal part where the LED lights of the photoelectric barriers were attached to the cage. After training was completed, I put a 25 cm x 100 cm piece of paperboard at the entrance of the corridor on the left side (bats´ perspective) leaving a 25-cm opening for the bats to fly through. In the second step, I put a second, similar sized, obstacle at the end of the corridor on the right side. To count the number of collisions with this second obstacle, I fixed a band of plastic foil at the edge of the paperboard with 2 cm wide cuts maintaining obstacle width to 25 cm. A movement of the plastic foil stripes due to a collision of the bats' wing tips was visible on the video recording as glimpses of reflection of IR-light. To measure flight speed I placed two photoelectric barriers at the entrance and exit of the corridor. The frontal infrared spot was focused on the plastic foil stripes of the second obstacle, whereas the other infrared spot pointed at the perch. Ten flights per trial from perch to feeding dish and ten flights back from the feeding dish to the perch were analysed from video recordings for each bat and each manoeuvre condition. I analysed and compared flight speed and number of collisions between three manoeuvre conditions for each bat and for each sample site.

2.3.2.6 Social interactions

The social behaviour of a group of maximally 22 individuals was synchronously audio and video taped in 2009/2010 inside the flight room where the bats were permanently kept (see Figure 3). Fifteen, respectively fourteen (one individual died) experimental individuals and seven additional individuals were kept as a stable social group. The seven additional non-experimental bats were females as I assumed from the literature that *Macrotus* males might occupy leks which are visited by several females. The group consisted of nine, respectively eight males and 13 females (see list of individuals kept in the lab Table 29 in the Appendix).

I observed for 1 up to 3 hours per session in the darkness phase of the bats, at variable daytimes. I focused on social interactions, like encounters of newly captured individuals and their release into the facility, allogrooming dyads and spatial allocation of the individuals.

Analysis was based on recordings done in the four months April to July 2010. I analysed ten days per months. Recordings were made before experiments/training on four days, immediately after the experiments/training on two days, and at least three hours after experiments on four days, in each month. I analysed one hour per days resulting in 40 hours. From these video sequences, spatial allocation, grooming events and contexts of social call emissions were analysed. Parts of the video analyses were done by F. Morazón within a MSc´s course.

2.3.2.6.1 Allogrooming partners

I recorded allogrooming events between two interaction partners. This very affiliative social behaviour was preferentially recorded whenever it occurred during the observation sessions of the social group. From the video, I determined individuals involved and their role in allogrooming as being the receiving bat or the grooming bat. The analysis was done on allogrooming dyads to see if there are preferences for specific individuals and/or locality members. Obtained frequency data was tested by calculating Chi^2-tests.

2.3.2.6.2 Social calls

Social behaviour was documented on video- and synchronously on high speed audio recordings in two situations: of 21 individuals from three samples sites inside the facility and of a small number of individuals (<10) from the same sample site inside a transportation cage, before they were released into the facility.
All audio recordings were checked for the occurrence of social calls of analyzable quality. Recordings were analysed whenever sample size, recording quality and unambiguity of individuals, context and behaviour were given. I measured syllable duration (ms), peak frequency of the fundamental (kHz), noted number of harmonics and time-frequency contour and based the analysis of context specific social calls on individuals from the three sample sites.

2.3.2.7 Spatial allocation

The positions of bats inside the flight room were determined every five minutes, beginning from the tenth minute of a recording hour. For this purpose, I defined fixed areas within the flight room (see Table 4 and Figure 27 in the Appendix).

I focussed on every individual while recording the video to later verify spoken comments about the position of every bat on the recording. Occasionally an individual was not identifiable on the video or could not be located. In this case the first available snap shot was analyzed instead. I determined the individuals´ position from the video recording and analysed if individuals had preferred areas within the room, if they had preferred neighbours, and/or if there was a preference to be close to locality mates. Resulting data were frequencies which I compared to a equal frequencies of occurrence in the areas and tested with Chi²-tests.

To test if time influenced the spatial allocation of bats inside the flight room, I used standardized frequencies of each bat at the areas and pooled five days. Resulting data basis consisted of eight repeated measurements per area and per individual which were used in the Friedman ANOVA.

Table 4: Defined areas in the flight room where bats were permanently housed. List refers to most frequently used perches. In addition, it was noted whether a bat was flying or not visible ("absent") in the respective video snapshot.

Area	Description
Wall	Area on the wall where the cable is fixed.
Cable 1	First meter of the cable until second bulb.
Cable 2	Second meter of the cable until third bulb.
Cable 3	Third meter of the cable until fourth bulb.
Roost	Upper corner to the left of the cable.
Edge	Edge on the wall where the cable was fixed, 15 cm below the cable fixation spot along the entire wall.
Wall 2	Area above the edge on the wall.
Ceiling	Opposite upper corner to the roost on the right side of the cable.
Feeding area	On the wall to the left of the cable. From about 150 cm height to the ground.
Flying	Bats were flying within the room.
Absent	Position of the bat was not observable.

2.3.3 Statistics for field studies and studies on behaviour in the lab

For calculations and statistics I used Excel (Microsoft Excel 2003), Statistica (6.1, StatSoft Inc., 1984-2004) and PASW Statistics (18.0.0, SPSS Inc. 1993-2007).

To see whether bats from the four sample sites differed acoustically and/or morphologically we included only individuals of which we had both data sets complete. Outliers and extremes in both data sets were excluded. Overall level of significance (α) was 5 %, except for pairwise comparisons, where the Bonferroni corrected level of significance (α_{bonf} = α divided by number of comparisons) was applied. Taking the sample sizes into account, I decided to use nonparametric statistics for descriptive statistics (medians, grand medians, interquartiles, non-outlier ranges). Basis for calculations was always the individual. Grand medians of sample sites, when calculated, consisted of medians per individual.

Prior to multiparametric statistics I checked parameters for normality (Shapiro Wilk´s tests), homogeneity of variances (Lavenes´ test and Hartley tests, calculated as variance ratio = largest variance/smallest variance) and correlation. I used Spearman's correlation coefficient, i.e. a coefficient of zero would mean "no correlation", a coefficient of one "complete correlation". I defined the critical value for excluding a parameter for multivariate statistics at $\geq$ +/-0.7 (Zöfel, 1988).

To account for different scales of the multiparametrical data sets, I used z-transformed values (computed as: std. score = (raw score - mean)/std. deviation). To test for differences between the sample sites based on individuals, I calculated Kruskal-Wallis H tests. If significant differences were detected, this test was followed by multiple pair wise comparisons to reveal which of the sites actually differed from another. I applied the Mann-Whitney U tests to test for differences in average ranks of two samples.

For analysing fur coloration types I did Chi^2-tests on frequency data to see whether distinct fur colour types occur more often or less often than expected if coloration would occur at equal rates in the four sampling areas.

Due to the low number of analysable trials in the experiment of passive listening, the frequency of an material (plastic, paper, metal) on a position (left, middle, right) was not equal (overview of analysed sessions Table 30 in the Appendix). To account for this unbalanced design, expected frequencies for each material on a position were adapted. I assumed an equal number of landings and approaches to the respective material and to positions.

Subsequent to pair wise comparisons, a classification of median echolocation pulses and morphological parameters according to the sample sites was achieved with MANOVAs. MANOVAs and subsequent discriminant analysis have assumptions which had to be checked. Multivariate normality, i.e. normal distribution within each group, has to be given, parameters must not highly correlate and homogeneity of covariances should be equal for each group. Morphological data were standardized into reciprocally transformed parameter values (computed as: 1/(variable value +1)) to reduce the effect of non-homogenous variances in some morphological parameters. MANOVAs test statistics Pillai´s Trace (if sample sized are equal), Wilk´s Lambda, Hotelling´s Trace and Roy´s Largest Root (not robust when homogeneity of variances is untenable) helped to decide whether the effect of sampling site on echolocation pulses and/or morphological parameters is significant or not. The univariate test statistics report the effect of sample site for each parameter separately. MANOVAs post-hoc tests, Gabriel (when sample sizes are different) and Games-Howell (samples differ in variances and sample sizes are unequal) were performed to explore the data for any between-group-difference and to check a priori predictions of groups according to our sample sites. In the discriminant analyses individuals were grouped to one of the four sample sites according to their individual multivariate parameter characteristics which were entered stepwise into the analyses (F in = 3.84 and F out = 2.71). For classifications, group size was taken into account, the leave-one-out method was applied and resulting values were cross-validated. The discrimination analysis allowed an evaluation of the discriminating power of each morphometrical parameter.

Furthermore, the comparison of predicted sample site membership, which is the site were the bat was caught, with assigned membership results in a measurement for morphological/acoustical distinctiveness between the four sites.

2.4 Genetic variability

I collected tissue samples from those individuals which were also captured for echolocation call recordings. In total, I sequenced twenty individuals of the Cuban *M. waterhousii*, five of each geographic sample site and added sequences downloaded (from GenBank) of related species as outgroups.

2.4.1 Gene sequencing procedures

Tissue samples were stored in 70% alcohol, later in a Lauryl-Sarcosin buffer. Prior to DNA extraction, tissue was digested using Proteinase-K (Sigma-Aldrich Co., USA).
Genomic DNA extraction followed a standard phenol-chloroform-isomethyl alcohol method. To extract DNA and remove protein contaminants we added 500 µl phenol-cholophorm-isomethyl alcohol, and centrifuged for 5 min at 13 000 rpm. Upper clear aqueous layer contained alcohol and DNA. After transfer of this layer into a new tube this step was repeated with 500 µl cholorophorm-isomethyl alcohol. Before purifying the DNA extract, I added 500 µl 100% ethanol and freezed the samples for 5 min at -80°C to precipitate DNA. Purifying was done twice by adding 70% ethanol and centrifuging at 13 000 rpm for 5 min to remove salts and small organic molecules. After drying the DNA pellet in a vacuum pump I re-suspended DNA in 20 µl Low TE to protect DNA from degradation by DNase and to keep the DNA stable in a pH buffered environment before running a Polymerase Chain Reaction (PCR).
Aliquotes of the DNA isolates were diluted if necessary to achieve a standard concentration of 40 ng/µl for PCR. Optimal DNA concentration and temperature

during annealing were ascertained by running gradients and comparing PCR products on the electropherogram.

To amplify two mitochondrial markers, Cytochrome-b (cyt-b) gene and the d-loop region (d-loop), I used primers which were developed and successfully applied in previous studies. Those were for a 482 bp fragment of the cyt-b gene: MVZ05 (L14724) forward 5´-CGA AGC TTG ATA TGA AAA ACC ATC GTT G-3´ (Kocher et al., 1989a; Smith & Patton, 1991; Van Den Bussche & Baker, 1993a; Dávalos & Jansa, 2004) and MVZ04 (H15149) reverse 5´-GCA GCC CCT CAG AAT GAT ATT TGT CCT C-3´ (Kocher et al., 1989b; Van Den Bussche & Baker, 1993b) and for the d-loop region fragment of 350 bp: F1 (forward 5´-CCC CAC CCT CAA CAC CCA AA-3´ (Pumo et al., 1988; Fleming et al., 2009) and F ((F-H) reverse 5´-GTT GCT GGT TTC ACG AG GTA G-3´ (Wilkinson & Chapman, 1991; Fleming & Murray, 2009; Fleming et al., 2009). Total PCR reaction volume was 25 µl per sample containing for cyt-b PCR 2.5 µl 10x advanced buffer, 0.5 µl dNTP´s 10 mM (5prime), 1 µl of each primer at 10 pmol/µl, 0.25 µl Taq (5prime) and 2.0 µl DNA. For d-loop PCR the reaction volume of 25 µl contained 2.5 µl 10x advanced buffer, 4.0 µl dNTP´s at 0.4 nM (5prime), 2.5 µl of each primer at 14 pmol, 0.4 µl Taq (5prime) and 2.0 µl DNA. Samples were entered into the Thermocycler and PCR run under the following thermal conditions: For cyt-b: 94°C for 90 sec (initial denaturation), followed by 35 cycles of 94°C for 60 sec (denaturation), 50°C for 60 sec (primer annealing) and, 72°C for 90 sec (elongation), and 72°C for 60 sec (final elongation). After successful termination I stored PCR products at 4°C overnight. For d-loop I started PCR at 94°C and 120 sec. In 30 cycles DNA was denaturated at 94°C for 40 sec followed by annealing at 55°C and 60 sec and by extending at 72°C for 20 sec. Final elongation was at 72°C for 300 sec.

To check if PCR was successful (length of fragment and amount of product) I run an agarose gel electrophoresis with each sample and a control without DNA. For gel electrophoresis 2 µl 6x Blue were mixed with 3 µl DNA and applied on the gel. Gel was run for 30 min with 95 V. Samples were exposed to UV in a Transilluminator and visible banding pattern was saved digitally as a electropherogram. Concentrations of PCR products were measured with a photometer to ensure the concentraion of at least 20 µg/µl as required for sequencing.

Prior to sequencing, PCR products were purified following manufacturers´ protocol (SeqLab two-step purifying kit). Sanger-sequencing was done by SeqLab, Sequence Laboratories Göttingen GmbH, Germany ("AdvantageRead").

Digitally received single sequence files contained forward and reverse strains and were manually corrected and saved as consensus sequences ("contigs") using the software SeqMan II (version 6.00 by DNASTAR Inc., Madison, WI, USA). Base-calling ambiguities between strands were resolved by choosing the call on the clearest strand. Sequences were automatically aligned following the Clustal algorithm by using the software MEGA4 (Tamura *et al.*, 2007b). The multiple sequence alignment was adjusted visually. I added sequences of *M. californicus*, *M. waterhousii* (from Mexico) and *Micronycteris megalotis* to the cyt-b alignment and *Phyllostomus hastatus* and *Uroderma magnirostrum* to both alignments (cyt-b and d-loop) which served as outgroups. These sequences were retrieved from GenBank (accession numbers are followed by respective authors and year of upload: *M. californicus*: AY380744.1, Porter, C. A., Hoffmann, F. G. and Baker, R. J. 2003; *M. waterhousii*: AY380745.1, Porter, C. A., Hoffmann, F. G. and Baker, R. J. 2003; *M. megalotis*: DQ077429.1, Porter, C. A., Hoofer, S. R., Cline, C. A., Hoffmann, F. G. and Baker, R. J., 2005; *P. hastatus*, cyt-b: FJ155479.1, Hoffmann, F. G., Hoofer, S. R. and Baker, R. J, 2008; d-loop: U95323.1, Wilkinson, G. S. and Mayer, F., 1997; *U. magnirostrum*, cyt-b: DQ312405.1, Hoofer, S. R. and Baker, R. J., 2005; d-loop: FJ154312.1, Velazco, P. M. and Patterson, B. D., 2008).

2.4.1.1 Sequence analyses

Both markers were analyzed independently. Descriptive information was obtained with Mega4 (Tamura *et al.*, 2007a). As a first step straightforward uncorrected p-distances were calculated. This measurement of genetic distances between sequences simply divides the number of non-identical nucleotide sites by sequence length, i.e. number of nucleotides. Sequence divergence is number of variable sites/sequence length*100. Changes of nucleotides can be caused by two substitution types, transversions (within chemical classes) or transitions (between chemical classes). If it leads to changes of amino acids they are non-synonymous substitutions. A high proportion of non-synonymous substitutions may indicate directional selection. To control for effects of multiple substitutions, distance corrections can be calculated and applied as best fit substitution models to each data set. For maximum parsimony (MP) and Maximum Likelihood (ML) analyses, PAUP* version 4.b10 (Swofford, 2003) was used. All sequences entered the analyses as single taxa and all characters entered the analyses with equal weighting. Gaps were treated as missing in neighbour-joining method (NJ) and ML, and as a fifths base for MP analyses. I retrieved informative data such as number of parsimony informative characters, genetic distances (p-values, gaps treated as missing value) and applied Neighbour-Joining method on distances to obtain a first tree. Analyses were conducted on informative positions only. I conducted heuristic searches with 100 replicates and randomly added sequences. The starting tree was obtained via stepwise addition of taxa which were added randomly. One tree was held after each addition. I chose tree-bisection-reconnection method as branch-swapping algorithm. To assess reliability of clades the number of bootstrap replicates was set to 1000. Results were displayed as bootstrap 50% majority-rule consensus trees. To chose the best fit substitution model for the Maximum Likelihood analyses I entered data sets into the software ModelTest version 3.7 (Posada & Crandall, 1998) to find the best model of nucleotide evolution for each data set. The Akaike information criterion and the hierarchical Likelihood Ratio test resulted in identical best fit models. They give information about the goodness of fit of the respective model to the data set. The model settings were transferred into PAUP*. The general time reversible (GTR) model best fitted the cyt-b data and the Hasegawa-Kishino-Yano model (HYK) best fitted the d-loop data. Both models account for unequal frequencies of the four nucleotides. The HKY model (Hasegawa *et al.*, 1985) accounts for unequal rates of transitions and transversions. The GTR model (Rodríguez *et al.*, 1990) assumes a symmetric substitution matrix (means that A changes to T with the same rate as T changes to A and so on) and specifies unequal rates among all pairs of substitutions.

I approached phylogenetic relationships by comparing the results as phylogenetic trees for Neighbor joining, Maximum Parsimony and Maximum Likelihood. For tree editing TREEVIEW version 1.6.6 (Page, 1996) was used.

In addition, population genetics were analyzed and haplotype networks generated. A haplotype network can depict relationships between sampled haplotypes especially, if sampling was done on an intraspecific level where the period of time in which genetic markers have evolved is too short. It visualises the number of mutational steps between the haplotypes. I constructed minimum spanning trees using HapStar (Teacher & Griffiths, 2011). Network approaches take into account several features associated with intraspecific gene evolution (Posada & Crandall, 1998). For population genetics Arlequin version 3.5.1.2 (Schneider *et al.*, 2000) was used. I either chose to analyze all consensus sequences as one population or as four populations referring to the four sample sites. Analysis of Molecular Variance (AMOVA) was calculated to reveal genetic structuring (Weir & Cockerham, 1984; Excoffier *et al.*, 1992; Weir, 1996; Posada & Crandall, 1998). The Mantel test was applied to test genetic isolation by geographic distance. For this test I standardized pairwise genetic distance data (F_{ST} [$F_{ST}/(1-F_{ST})$]) (Rousset, 1997). I analyzed the correlation between two matrices using standardized values of pairwise genetic differentiation estimates versus direct aerial distances. If the resulting test statistic g > critical value, test is significant means there is a correlation.

For the complete list of standard molecular diversity indices and population genetic parameters please see following table (Table 5).

Table 5: Standard molecular diversity and population genetics measurements calculated for cyt-b sequences and d-loop sequences.

Parameter	Definition
S	Number polymorphic sites.
TI	Number of observed transitions.
TV	Number of observed transversions.
π	Nucleotide diversity, i.e. sequence diversity.
h	Haplotype diversity.
N_h	Number of haplotypes.
p-dist	Uncorrected p-distances: genetic distance, based on genetic polymorphism data, such as single-nucleotide polymorphisms (SNPs).
Pi	Mean number of pairwise differences.
Tajima´s D	Tests of selective neutrality.
AMOVA	Analysis of Molecular Variance. Genetic structure analyses.
Global F_{ST}	Fixation index (FST) based on pairwise genetic distances, is a measure of population differentiation. It reflects the degree of differentiation among all sample sites relative to random haplotypes drawn from the entire population.

3. Results

3.1 General observations on Cuban *M. waterhousii*

I made some general observations during the field trips about roost usage, diet, colony size and reproduction cycle which I will briefly summarize here.

I found *M. waterhousii* in small caves as well as in large and deep caves. It was not restricted to areas of total darkness and occurred in illuminated zones as well as in dark chambers. The species often roosted in parts of caves with a height >3 m where it preferred ventilated areas, confirming earlier observations. Night roosts were not identical with day roosts. Night roosts were regularly visited by the same individuals as I recaptured marked individuals when I re-sampled a cave in subsequent nights. At night roosts, captured prey was eaten at constant perches as revealed by accumulations of insect parts on the ground (mainly wings). The diet consisted primarily of insects; occasionally small vertebrates such as lizards were caught. New born mice were taken immediately by bats in captivity. In contrast to reports in the literature, I could never observe *M. waterhousii* capturing prey on the wing, neither in nature nor captivity. It gleaned prey from a surface and easily lifted up with prey directly from the ground. Hovering occurred regularly and in general flight was rather slow and highly manoeuvrable. Flight speed was higher in enclosed areas where the bats had to manoeuvre faster. I observed variable group sizes in day roosts throughout the year ranging from three, up to several hundred individuals. Gender ratio was also very variable throughout the year. I found mixed gender groups, as well as maternity colonies with a small proportion of males, pure male colonies and male colonies with a few females. Both, colony size and gender composition, may vary seasonally with reproductive cycle.

Even with taking the lack of data for some months into account, the data collected on reproductive status of 66 adult females and 78 adult males, allow to characterize the reproductive cycle of the Cuban *M. waterhousii* which matches previous findings from Cuba, Mexico, Haiti and even California (Silva Taboada,

1979). The only incongruent data on this reproductive cycle regards to Jamaican *M. waterhousii* which gave birth in December (Dávalos & Eriksson, 2003). Females captured in January in the western, central and eastern regions showed no sign of pregnancy. At the end of February until end of March, females were in the last stage of pregnancy. In May and June, females gave birth and were lactating. In July, most females were lactating (see Figure 25 in the Appendix) or already post-lactating. In September, females seem to almost complete their reproductive cycle as they were all post-lactating. To summarize, birth to offspring was given in April or May. Accordingly, in June, females still nursed their offspring. Juveniles should be weaned between July and September as I found more post-lactating females in these months. Thus, I assume the end/start period of the reproductive cycle at the end of the year.

In all sampling months, i.e. in January, February, May, July, and September, I caught males with secreting head glandulae (see Figure 26 in the Appendix). There was no difference between the four geographical sites concerning season for developing secreting glandulae, but there was a seasonal difference in the secretion activity and the number of males with no secreting glandulae; while males captured at the beginning of the year in the western, central, and eastern region mostly had no secreting glandulae and the fur was about to cover the bold part of the head again, all males on the offshore island, had well developed secreting glandulae. In relation to the female reproductive cycle I conclude, that the expression of secreting glandulae of males starts when juveniles became more independent in July to September. The months of October, November and December may therefore be the season of advertisement and mating and glandulae secretion might play a role in social interactions during the reproductive season of *M. waterhousii*. During the phase before mating, in a period where females were busy with nursing and still roosted in maternity colonies and males formed male day roost colonies, secretion of glandulae reached a peak of activity and males were bold-headed. As most of the captured males had no pronounced glandulae during the time of fertilization (in January), a stronger importance of glandulae secretion for male-male interactions is likely (Rehorek *et al.*, 2010).

3.2 Morphology

3.2.1 Description and comparison of morphometrical data of Cuban *M. waterhousii* from four sample sites

In general, the Cuban *M. waterhousii* is a medium-sized bat with an average forearm length of 52 mm and a wing span of about 31 cm. It has the habitus of a typical gleaning phyllostomid bat. It has an erected, almost triangular nose leaf of 8 mm length and prominent large ears of 25 mm length. The tragus is about three times shorter than the ear in length, acuminating to tip. The ears are joint at the basis, and the lower lip is furcated (for average measurements of morphological parameters, see descriptive statistics for morphometric measurements Table 31 in the Appendix).

3.2.1.1 Fur coloration

Concerning a geographic variation in fur coloration which was described for different morphotypes across the Caribbean distribution range of *M. waterhousii* (see introduction), I compared colorations between west, the offshore island, central and east Cuba. The 115 adult *M. waterhousii* specimens caught during field trips occurred in three major fur coloration types; brownish, reddish and greyish. The individuals varied in their coloration: 41.74% were brownish, 39.13% were reddish, and 19.13% greyish, whereas the greyish type occurred less often than expected, i.e. given the assumption that coloration types occur at equal frequencies. Compared to expected frequencies, observed frequencies of coloration types differed between the sample sites. On the offshore island, captured individuals were significantly more often greyish than expected. The reddish coloration type occurred more often than expected in the central region, whereas the greyish type occurred less frequently than expected. In the east, the brownish coloration type occurred more often than expected. In the west, observed frequencies of coloration types matched the expected frequencies, i.e. coloration types occurred at equal frequencies (Table 6).

Table 6: Coloration types and their frequencies at sample sites. Given are observed and expected frequencies and their degree of deviation. At 6 degrees of freedom significant results are marked with asterisks, whereas number of asterisks indicate level of significance (at 5%-level of significance the chi² value was 12.592 (*), at 1% 16.812 (**) and 0.1% 22.458 (***)).

	West	Offshore island	Central	East	sum
Observed					
Brownish	13	4	17	14	48
Reddish	8	2	27	8	45
Greyish	6	10	1	5	22
sum	27	16	45	27	115
Expected					
Brownish	11.27	6.68	18.78	11.27	
Reddish	10.57	6.26	17.61	10.57	
Greyish	5.17	3.06	8.61	5.17	
Degree of deviation					
Brownish	0.27	1.075	0.169	7.453 **	
Reddish	0.63	2.899	5.007 *	0.625	
Greyish	0.133	13.386 ***	6.726 **	0.006	
sum					31.666

Besides the prominent head glandulae characteristic of males, I found a sex dimorphism in fur coloration which was mentioned in previous studies. Due to the unequal number of captured males and females (proportions: all: 6:4, west: 6:4, island: 2:8, central: 7:3 and east: 7:3) and a low number of observations, no multiple Chi² test was calculated taking sex, coloration type, and geographical site into account. Therefore, I tested for a gender bias and pooled males, respectively females, of four sample sites (see Table 7).

Table 7: Sexual dimorphism of fur coloration. Given are observed and expected frequencies and degree of deviation for for each of the three coloration types of males and females. At 5 degrees of freedom significant results are marked with asterisks, whereas number of asterisks indicate level of significance (at 5%-level of significance the Chi^2 value was 11.070 (*), at 1% 15.086 (**) and 0.1% 20.515 (***)).

Sex and coloration type	Observed	Expected	Degree of deviation
Male brownish	26	19.17	2.43
Female brownish	22	19.17	0.42
Male reddish	43	19.17	29.62 ***
Female reddish	2	19.17	15.38 ***
Male grayish	1	19.17	17.22 ***
Female grayish	21	19.17	0.18
sum	115	115	65.25

With exception of the brownish coloration type, which was almost equally present in both genders, sex dimorphism was found in the frequencies of reddish and greyish individuals; male reddish individuals occurred significantly more often than expected and reddish females were significantly too rare. Greyish males were significantly less often than expected, whereas number of greyish females matched expectations.

3.2.1.2 Multiparametrical analyses of morphometry

As matching data bases of morphometrical and echolocation data sets were required, I based multiparametrical analyses on the same individuals as analyzed in echolocation analyses (next paragraph) caught in the field. Thus, I analyzed 77 individuals in total, 15 from the west (10 males, 5 females), 33 individuals from the central region (22,11), 13 individuals from the offshore island (2,11) and 16 individuals from the east (10,6).

Correlations within the morphological parameter set were moderate, with the highest correlation between forearm length and ear length (Spearman´s $\rho = 0.62$,

for a cross-table of correlations, see correlation between morphometrical parameters Table 32 in the Appendix).
To inspect the influence of geographical longitude on morphological measurements I plotted each morphological parameter versus longitude of sampled caves (Figure 6). A gradual increase was observed for forearm length and ear length, but correlation was moderate (Spearman´s ρ =0.35, r^2 = 0.24 and 0.33, 0.13, respectively).

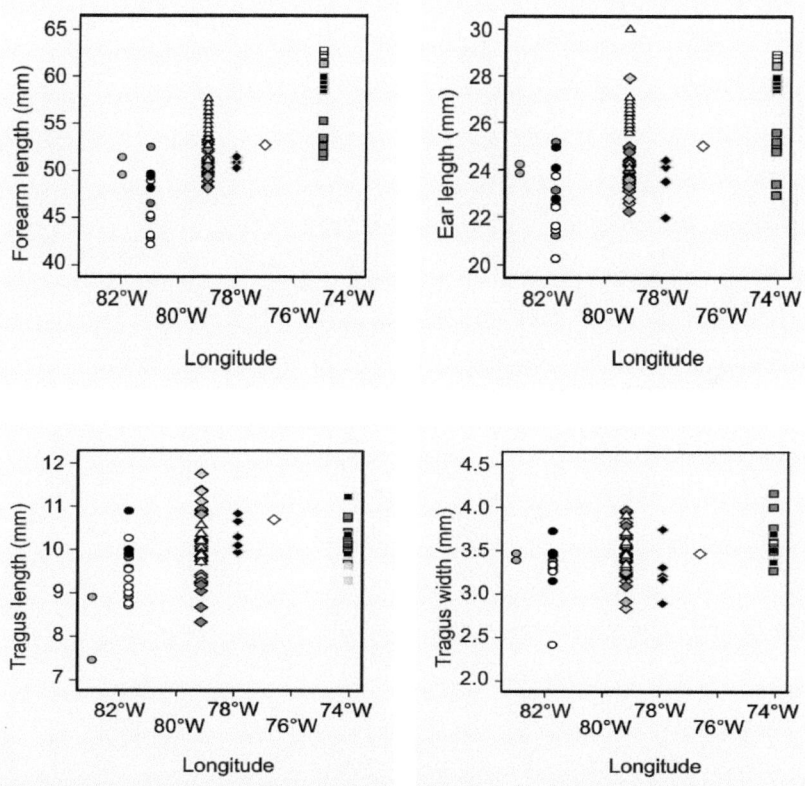

Figure 6: Morphometric parameters plotted as a function of longitude. Median values of every sampled cave complex are given. Sample sites are represented by a symbol and cave complexes by coloration (legend next page). Figure continued on next page.

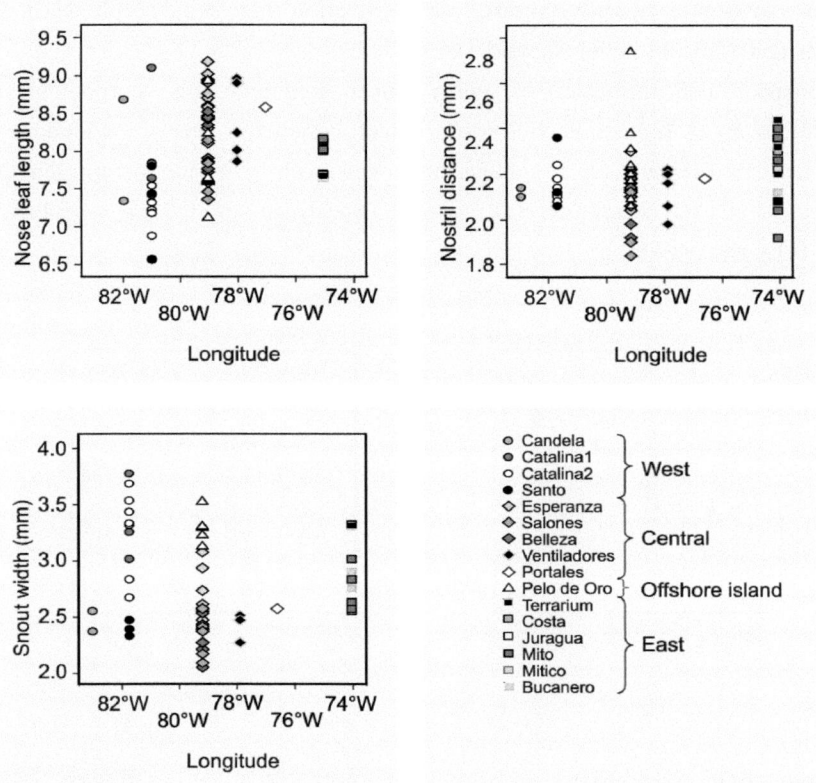

Figure 6 continued.

Prior to MANOVA and discriminant function analysis I checked the respective assumptions and found that normality was tenable for the eastern group in the parameters forearm length, nose leaf length and snout width (Shapiro-Wilk´s W = 0.876, p = 0.034, W = 0.778, p = 0.001, W = 0.784, p = 0.002). Whereas deviations from normality were caused by single individuals in the other groups, deviating distribution for the eastern group indicated binomial distribution shapes. Covariances of the parameters were not homogenous between the sample sites in nose leaf length, tragus length and snout width (Lavene´s test d.f. 1 = 3 and d.f. 2 = 73. F = 2.76, F = 4.71, F = 22.19, p = < 0.05). For statistical analyses, forearm length, ear length, nose leaf length, nostril distance, tragus length, tragus width and snout width were included. Descriptive statistics for each sample site and each parameter are presented as boxplots in Figure 7 (for a table giving medians and interquartiles, see descriptive statistics for morphometric measurements Table 31 in the Appendix).

The four sample sites differed significantly in all seven parameters except for nostril distance and tragus width (Kruskal Wallis test, p = <0.05, Table 33 in the Appendix).

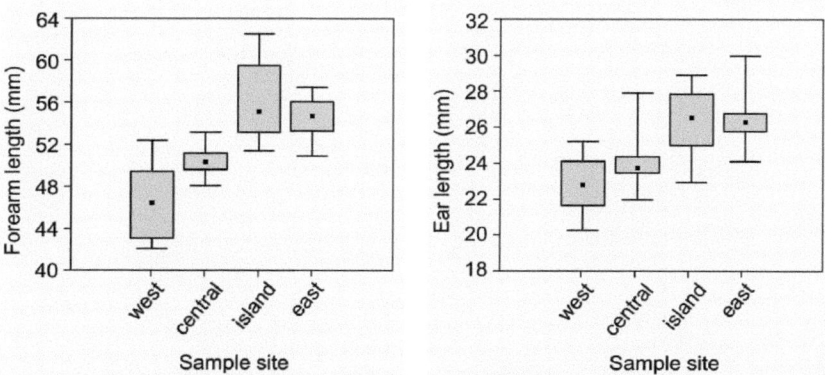

Figure 7: Morphometric parameters of bats from the four sample sites. Given are boxplots presenting median, and interquartiles and non-outlier ranges per sample site and parameter. Figure continued on next page.

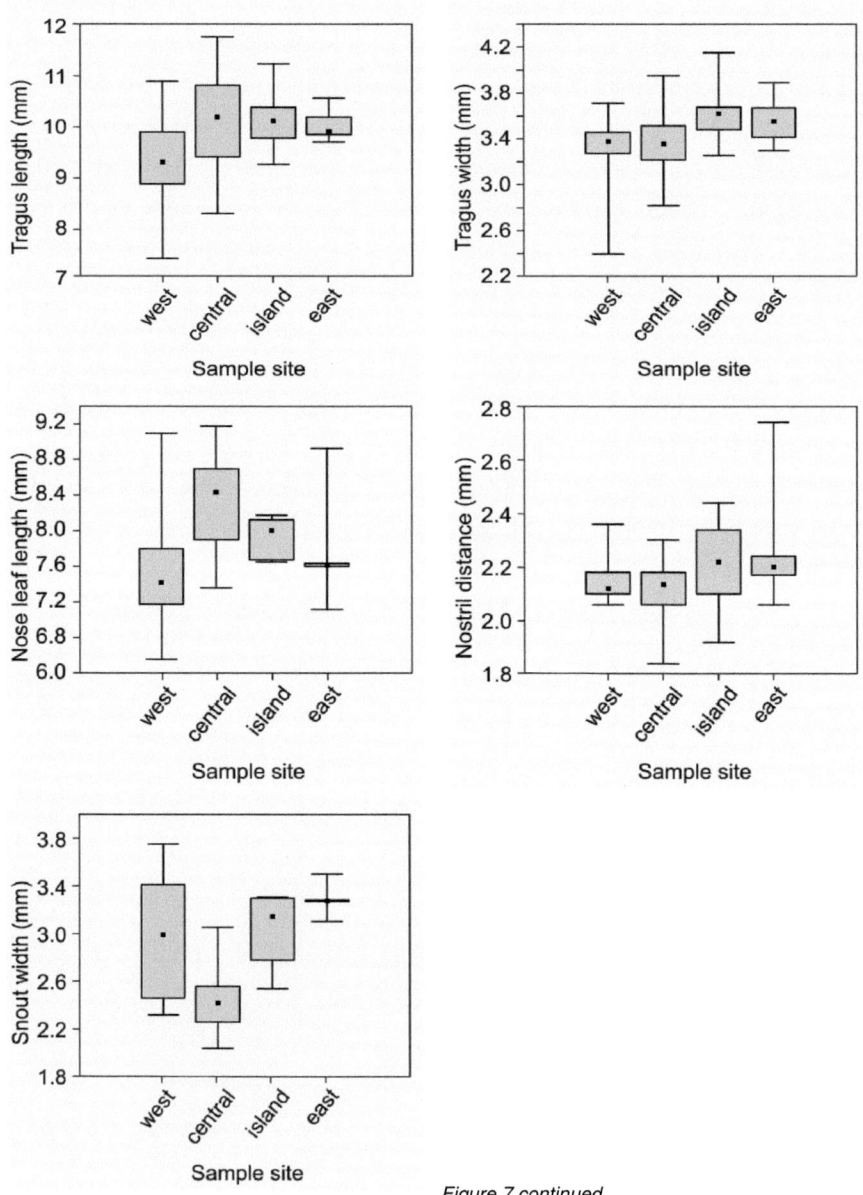

Figure 7 continued.

The MANOVA multivariate test statistic revealed an significant effect of site on the morphometric parameters (Wilk´s lambda (d.f. 1 = 21, d.f. 2 = 193) F = 12.00, p = 0.000 and Pillai´s trace (d.f. 1 = 21, d.f. 2 = 207) F = 8.73, p = 0.000). Nonetheless, no uniform pattern of morphometrical differences between the sites was obvious.

Parameters forearm length and ear length increased in size from west to east. However, individuals from the offshore island and from the east were generally similar. Although western individuals always had smallest values of median morphometric measurements, there was no congruent trend of increasing size eastwards.

Pairwise comparisons of sites, confirming the visual impression of the boxplots, revealed significant differences in several parameters. Whereas eastern and offshore island individuals differed only in the length of the nose leaf (7.62 to 8.08 mm) and in snout width (2.78 to 3.28 mm), all other sample site comparisons revealed three to four morphological parameters in different compositions in which sample sites differed significantly. Western individuals were significantly smaller than bats from the other three sample sites in five parameters (forearm, nose leaf, tragus and ear length), but not in snout width which was significantly larger in comparison to offshore island bats and not significantly different in comparisons to central and eastern bats. Surprisingly, individuals from the geographically closest sample sites, offshore island and central region, also differed in their morphology except for tragus length (Mann Whitney U test, p = <0.008, Table 8 and Table 34 for full test statistics in the Appendix. MANOVA post-hoc tests Gabriels´ and Games-Howell tests, p = <0.008, Table 8 and Table 35 for full test statistics in the Appendix).

Table 8: Results of multiparametrical pair wise comparisons calculated for each morphological parameter. Significant results are marked by "*", if Mann Whitney U test result was significant and by "+" if MANOVAs post-hoc tests were significant (Gabriel's and Games-Howell).

Parameter	West vs. Central	West vs. Offshore island	West vs. East	Central vs. Offshore island	Central vs. East	Offshore island vs. East
Forearm length	* |+|+	* |+|+	* |+|+	* |+|+	* |+|+	
Nose leaf length	* |+|+		* | |	* |+|+	| +	* | |
Nostril distance	* | |		* | |	* | |	* | |+	
Tragus length	* |+|+	* |+|+	* |+|+			
Tragus width			* | |+			
Snout width	* |+|+			* |+|+	* |+|+	|+|+
Ear length		* |+|+	* |+|+	* |+|+	* |+|+	

Although bats from the four sample sites differed across a combination of parameters, as considered in the MANOVA, I report here the summary of separate univariate ANOVA (d. f. 1 = 3, d. f. 2 =73) revealed that all parameters were affected by the parameter site (forearm length (F = 43.76, p= 0.000), nose leaf length (F = 12.27, p = 0.000), nostril distance (F = 4.90, p = 0.004), tragus length (F = 5.024, p = 0.003), tragus width (F = 4.92, p = 0.004), ear length (F = 24.77, p = 0.000) and snout width (F = 32.58, p = 0.000)).

The discrimination analysis allowed an evaluation of the discriminating power of each morphometrical parameter and classified each individual to one of the sample sites based on its morphometrical characteristics. The overall classification success achieved 70% for cross validated data (see classification of individuals based on morphology Table 36 in the Appendix). Classification revealed a grouping of individuals from same sample site especially of central individuals of which 97% were correctly classified. Only one central individual was falsely assigned to the western sample site. One third of western individuals were morphologically more similar to central individuals and one individual was assigned to the offshore island sample site. A high correct classification rate was also found for offshore island bats; 84.6% were correctly classified and only one falsely attributed to the western and to the central sample site each. Most misclassifications occurred for individuals from the east. More than half of the sample was assigned to the central site, 30% to the offshore island and only 12.5% to the predicted eastern sample site. This grouping was reflected in a plot of individuals placed in the space spanning the two first canonical functions (Figure 8) which explained in combination 99.4% of the variance (function 1: 53.6%, Eigenvalue = 1.48, canonical correlation = 0.773, Wilk´s lambda = 0.175, p = 0.000. Function 2: 45.8%, Eigenvalue = 1.26, canonical correlation = 0.747, Wilk´s lambda = 0.435, p = 0.000).

The functions consisted of all morphometrical parameters but with different discriminating power each. The contribution of the parameters to the two discriminating functions was different; in function 1, snout width and nose leaf length were the parameters contributing most (canonical correlation coefficients of $r = 0.793$ and $r = -0.566$ respectively). In function 2, forearm length, ear length and snout width had the highest discrimination power ($r = 0.997$, $r = 0.535$ and $r = 0.339$). Lowest contribution to sample site discrimination had nostril distance and tragus length ($r = < +/- 0.09$).

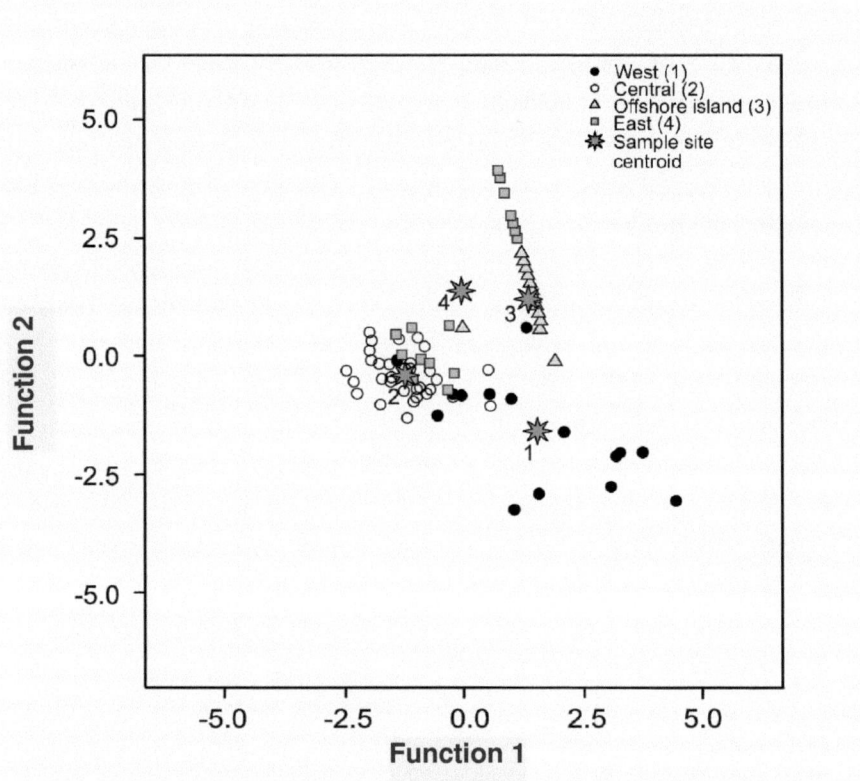

Figure 8: Plot of first two canonical functions that resulted from a discriminant analysis of multiple morphological parameters. Measures were taken from 77 adult M. waterhousii from four sample sites on Cuba. Each symbol refers to one sample site.

3.2.1.2.1 Summary

My study of the morphology of Cuban *Macrotus* revealed a sex dimorphism in fur coloration and the occurrence of head glandulae. Head glandulae occurred only in males and might play a role in male-male interactions during the mating season. From the three coloration types, most males were reddish, most females greyish and the brownish type was found in both sexes equally often. Morphometric measurements did not confirm the existence of two morphotypes of different size on Cuba. Rather, the size (i.e. forearm length) of individuals on the Cuban mainland was gradually increasing from west to east. However, the inclusion of other morphometric parameters showed an incongruent pattern of variability across Cuba, thus contradicting a cline. The type of parameters which separated the sample sites from each other varied between the pair wise comparisons. Overall, most discriminating were the parameters snout width, nose leaf length, forearm and ear length. Differentiation of samples sites was not associated to geographic distance, and most similar were individuals from the east and the offshore island. Most different were central individuals from the other sample sites.

3.2.2 Manoeuvrability

This experiment took place in the experimental cage in the lab (section 2.3.2.5).
In twelve of the fourteen tested bats (four western, 5 central, 5 eastern), flight speed differed significantly between the three manoeuvre conditions (Friedman ANOVA, d.f. = 2, p = 0.02). Subsequent pair wise comparisons between the conditions revealed that none of the bats increased flight speed when they had to manoeuvre around a single obstacle compared to a flight through the corridor without any obstacle. However, in the two obstacle condition, all bats generally increased flight speed compared to the condition with no obstacle from 0.92 m/s in average to 1.15 m/s in average (see Figure 9). Six bats increased flight speed significantly between the one and the two obstacle condition (Table 37 in the Appendix).
Four of five central individuals and two of five eastern individuals increased flight speed significantly, whereas none of the western individuals showed a significant increase of flight speed. The increasing difficulty in the two obstacle condition was also reflected in the number of collisions of the bats with the obstacles. Whereas no bat collided with the single obstacle, all bats collided with

the second obstacle in the two obstacle condition (see Table 9). Number of collisions varied between the individuals, but also between the sample sites. Western individuals collided on average 13 times in 80 passages (16%), while central individuals collided 54 times in 100 passages and eastern individuals even 66 times in 100 passages. Pair wise comparisons of number of collisions between western and central, as well as western and eastern bats were significantly different (Table 9, lower part).

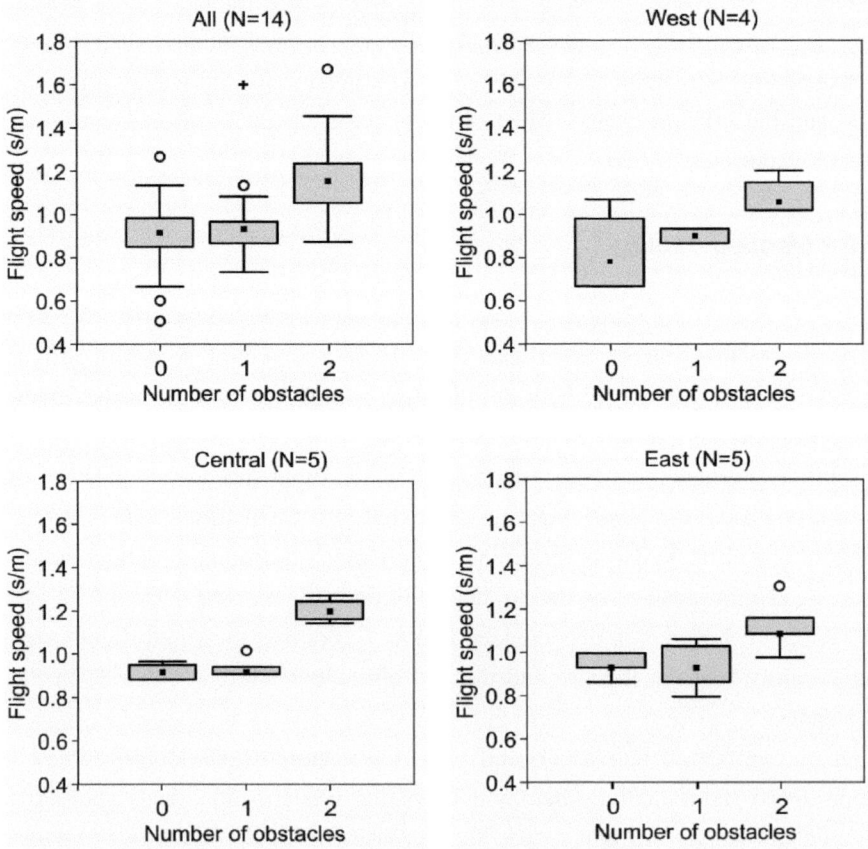

Figure 9: Flight speed under three experimental conditions. Boxplots represent median and interquartiles in the presence of, 0 = no obstacle, 1 = one obstacle and 2 = two obstacles for all fourteen individuals, and for western, central and eastern individuals.

Table 9: Number of collisions of the bats with the second obstacle in the two-obstacle flight course. Given are total numbers. Statistical sample site comparison is given in the lower part of the table.

Sample site	Bat	Total number of collisions (flight from perch, flight to perch)
	A	3 (1, 2)
	B	3 (1, 2)
	E	2 (2, 0)
	H	5 (1, 4)
West		13
	J	15 (6, 9)
	K	6 (2, 4)
	N	9 (5, 4)
	O	10 (5, 5)
	P	13 (6, 7)
Central		54
	S	12 (4, 8)
	T	11 (5, 6)
	V	12 (6, 6)
	X	16 (7, 9)
	Z	15 (6, 9)
East		66
Sample site comparison		Kruskal-Wallis H (d.f. = 2, N = 14) = 8.961 $p = 0.011$
Rank Sums		10.0, 35.0
U		0.000
Z		-2.470
p-level		0.014

3.2.2.1 Summary

All bats collided with the obstacle in the two obstacle condition, but bats from the western sample site had fewest collisions, bats from the central sample site and the east collided more often whereas bats from the east had most collisions. Although all bats had to manoeuvre to enter and exit the corridor (25 cm width), wing span below 30 cm allowed to pass without an increase in flight speed, whereas bats with a wing span above 30 cm had to increase flight speed to manoeuvre around the two obstacles, but still collided with obstacles often. Western bats were the smallest individuals with a wing span of 28.8 cm. Bats from the central region had a wing span of 31.3 cm and bats from the east 33.1 cm.

3.2.3 Summary of morphology

My findings did not confirm the existence of two morphotypes of Macrotus on Cuba. There was no abrupt size difference of *M. waterhousii* across Cuba, although individuals from the west were on average smaller than the others. However, the analyses revealed a structure of morphological variability across Cuba: individuals from the east and the offshore island were large in size and morphologically similar. Although geographically very close to each other, individuals from the offshore island and the central region were very distinct. Individuals from the central area did not represent an intermediate form between western and eastern bats. Morphologically most similar to the central individuals were western individuals.

The pattern of morphological characteristics was not highly correlated with longitude.

The influence of morphology on manoeuvrability was apparent in the obstacle experiment. Increased flight speed and number of collisions reflected differences in size, i.e. in wing span. The smallest individuals had no problems manoeuvring around two successively arranged obstacles, but larger individuals collided significantly more often. Smaller individuals also did not significantly increase flight speed with increased manoeuvre demands, but some of the larger individuals did. Larger individuals had to manoeuvre more, i.e. had a higher horizontal lean angle of the body, to pass through the 25-cm entrance/exit. As a consequence, they might have to increase flight speed to maintain uplift while manoeuvring which would cause slower speed in larger (thus heavier) bats. Large bats which did not increase flight speed collided with the plastic stripes of the obstacle, but managed to have fewer collisions than the similar sized bats. Those individuals were versed in all experiments, behaved calmly and were used as tutors for hesitating individuals.

3.3 Acoustic orientation and foraging behaviour

3.3.1 Description and comparison of orientation pulses of Cuban *M. waterhousii* from four sample sites

The echolocation pulse design of the Cuban *M. waterhousii* is typical for a phyllostomid bat which forages for insect prey close to the ground within the vegetation. Figure 10 shows an echolocation pulse of short duration, multiharmonic structure, broad bandwidth and high frequency.

The short frequency downward modulated echolocation pulses (0.9 ms of duration) consisted of up to five harmonics and covered a frequency range of 81 kHz with a peak of intensity at 79 kHz. Modulation rate was 0.081 kHz/ms and pulses started in median at 110 kHz, median lowest frequency was 33 kHz (for descriptive statistics of echolocation pulse measurements Table 38 in the Appendix).

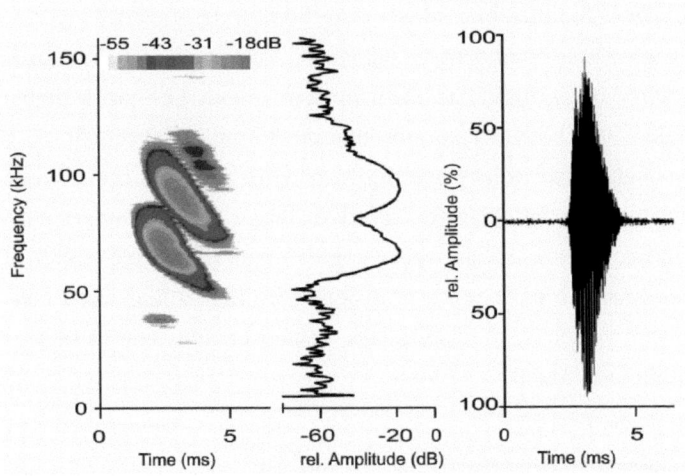

Figure 10: A typical echolocation pulse emitted in a standardized orientation situation in the field by a Cuban M. waterhousii. Given are sonagram, power spectrum and oscillogram.

3.3.1.1 Multiparametrical analysis of echolocation pulses

I analyzed echolocation pulses of 87 individuals of which I obtained both, morphological data and acoustic recordings. For multivariate analyses, I excluded outliers and extremes, thus achieving a data set of 77 individuals in total. Fifteen individuals were from the west (10,5), 33 individuals from the central region (22,11), 13 individuals from the offshore island (2,11), and 16 individuals from the east (10,6).

As body size can affect acoustic characteristics of echolocation pulses, I analysed allometric relationships between the matching data sets of acoustics and morphology. There was a negative correlation with forearm length, and ear length. Terminal frequency of the second harmonic was higher, if forearm length and ear length were smaller (Spearman´s $\rho = -0.7$, for both, see Table 39 in the Appendix).

Correlations within the acoustic parameter set did occur, too, and to reduce redundancy effects, I excluded highly correlating (Spearman´s $\rho = > +/-0.7$) parameters. Thus, the remaining parameter set for analyses, consisted of pulse duration, peak frequency, terminal frequency, effective bandwidth of the pulse and peak frequency and terminal frequency of the second harmonic.

Effects of longitude on the acoustic characteristics were not evident by correlation analysis (highest value Spearman´s $\rho = -0.35$ for peak frequency).

For MANOVA and discriminant analysis, I assumed four acoustic groups, a western, a central, an offshore island and an eastern group. In the sample of the central group, peak frequency of the pulse, terminal frequency of the pulse and effective bandwidth deviated significantly from a normal distribution (Shapiro-Wilk´s $W = 0.883$ and $p = 0.002$; 0.715 and $p = 0.000$; 0.920 and $p = 0.018$). As this was due to the occurrence of the fundamental, I retained these parameters in the set as important characteristics of echolocation pulses of individuals from this sample site. The parameters of echolocation pulses from eastern bats were normally distributed, except for the terminal frequency of the pulse (Shapiro-Wilk´s $W = 0.848$ and $p = 0.013$). Again, I did not exclude terminal frequency from the parameter set as its deviation from a normal distribution is also ascribed to the emission or not emission of the fundamental and therefore is a meaningful characteristic. The tests for homogeneity of variance for the four groups revealed that only effective bandwidth had unequal variances (Lavene´s $F = 7.192$).

Descriptive statistics of acoustic parameters for the four sample sites are presented as boxplots in Figure 11.

Some of the differences in acoustic structure of the echolocation pulses may be a result of the presence of the fundamental. It was not emitted by all individuals and in addition it was of variable intensity in those pulses with a fundamental (Table 10). All western bats emitted intense fundamentals, as well as central individuals. In pulses of eastern bats, the fundamental was emitted, but mostly with low intensity. Central bats mostly emitted pulses without a fundamental and if it was emitted, it was faint.

Table 10: Overview of emission of the fundamental in echolocation pulses. Given are absolute numbers representing numbers of individuals. A fundamental was absent if no indications from visual inspection of sonogram and no peak in the power spectrum was measurable.

Sample site	No fundamental	With fundamental	Fundamental not in effective bandwidth
West (N=20)	0	20	2
Central (N=34)	0	34	5
Offshore island (N=15)	10	5	5
East (N=18)	0	18	8

The statistical comparisons of acoustic characteristics of the four sample sites revealed significant differences in each parameter (Kruskal Wallis H test Table 40 in the Appendix) and the MANOVA revealed a significant effect of sample site on the acoustic structure (Pillai´s trace (d. f. 1 = 18, d. f. 2 = 210) $F = 13.665$, $p = 0.000$, Wilk's lambda (d. f. 1 = 18, d. f. 2 = 193) $F = 7.864$, $p = 0.000$) and separate univariate ANOVAs (d. f. 1 = 3, d. f. 2 = 73) revealed that sample site affected all six acoustic parameters; pulse duration ($F = 7.488$, $p = 0.000$), peak frequency of the pulse ($F = 6.039$, $p = 0.001$), terminal frequency of the pulse ($F = 159.851$, $p = 0.000$), effective bandwidth ($F = 5.378$, $p = 0.002$), peak frequency of the second harmonic ($F = 20.195$, $p = 0.000$) and terminal frequency of the second harmonic ($F = 38.995$, $p = 0.000$). Ranges of acoustic parameters compared between the four sample sites overlapped in most parameters, but not between all four sample sites. Two main patterns were obvious from the boxplots; terminal frequency and bandwidth, as well as peak-, start-, terminal frequency and effective bandwidth of the second harmonic showed a grouping of pulses from individuals from western-central sample sites and of offshore island-eastern sample sites. The parameters pulse duration, start frequency and modulation rate did not show a remarkable differences between the sample sites.

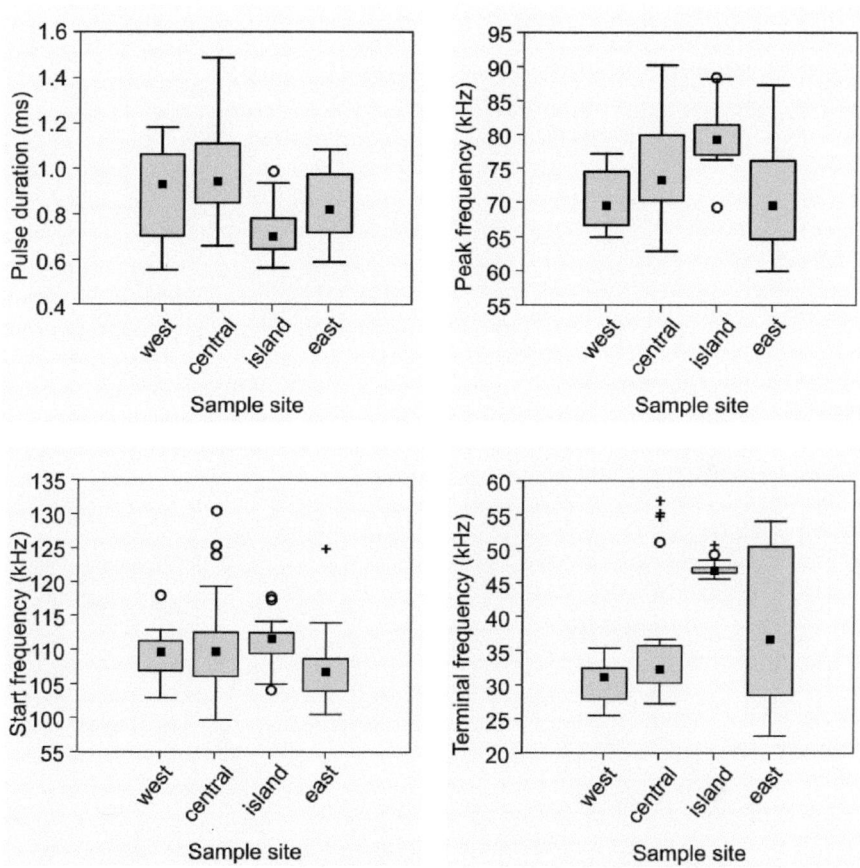

Figure 11: Acoustic parameters of bats from four sample sites. Box-whisker plots show Grand medians, upper and lower quartiles, non-outlier ranges, outliers and extremes per site for each acoustic parameter. Please note that statistical testing was based on data excluding outliers and extremes. Figure continued on next page.

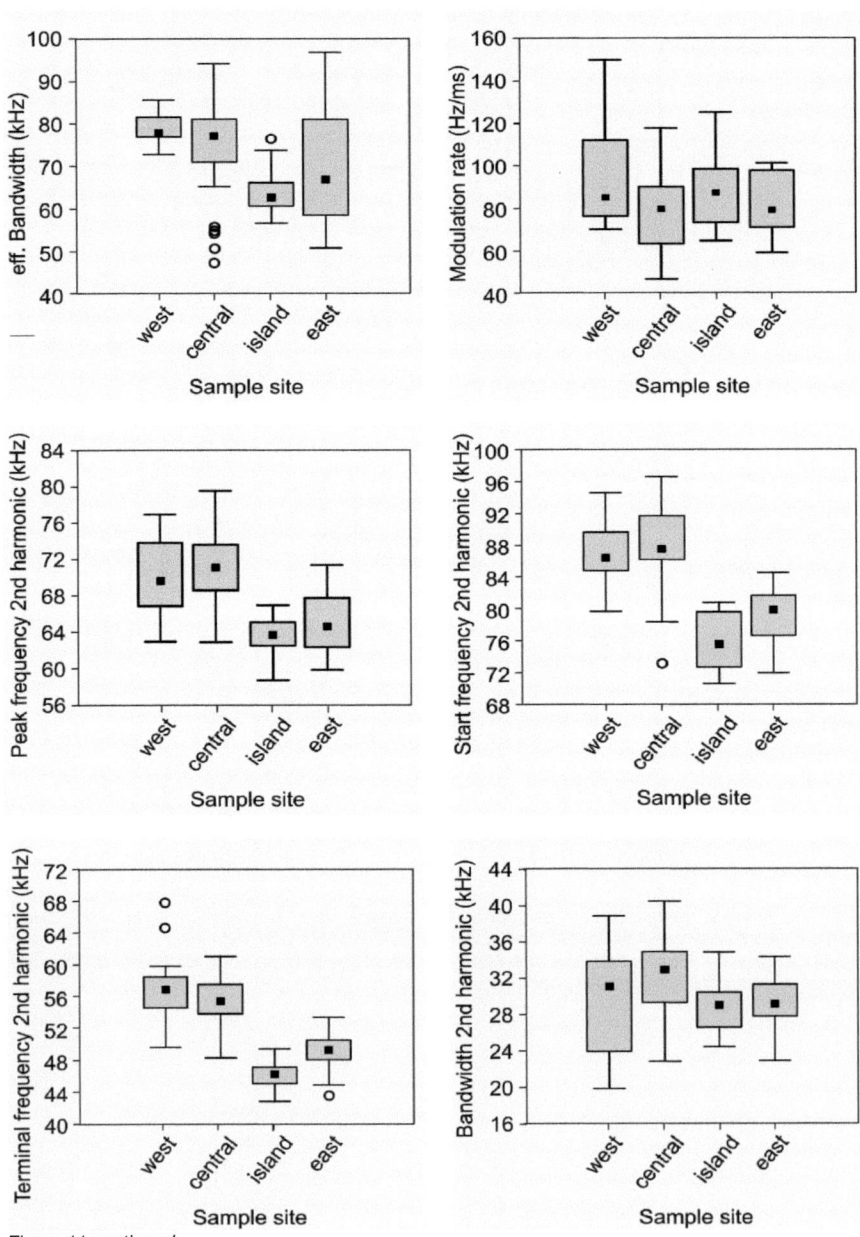

Figure 11 continued.

Subsequent multiple pair wise comparisons revealed that median pulses of western individuals and central individuals were acoustically indistinguishable. Table 11 summarizes statistical results of sample site comparisons (Mann Whitney U tests Table 41 in the Appendix and Gabriel's and Games-Howell tests as post hoc tests for MANOVA Table 42 in the Appendix).

Table 11: Results of multiparametrical pair wise comparisons calculated for each acoustic parameter. Significant results are marked by "*", if Mann Whitney U test result was significant and by "+" if MANOVAs post-hoc tests were significant (Gabriel's and Games-Howell).

Parameter	West vs. Central	West vs. Offshore island	West vs. East	Central vs. Offshore island	Central vs. East	Offshore island vs. East
DurP		|+|		*| |	| |+	
PeakfP		*|+|+				*|+|+
TermfP		*|+|+		*|+|+	|+|+	|+|+
eff.Bandw		*|+|+		*|+|+		
2ndHarmPeakf		*|+|+	*|+|	*|+|+	*|+|+	
2ndHarmTermf		*|+|+	*|+|+	*|+|+	*|+|+	*|+|+

The median pulses of western and offshore island individuals highly differed. Median pulses of western individuals were longer in duration (0.89 ms versus 0.71 ms), lower in peak frequency (71.9 kHz versus 79.2 kHz), lower in terminal frequency (30.8 kHz versus 47.0 kHz), and had a broader effective bandwidth (78.0 kHz versus 61.9 kHz). In the two frequency measurements of the second harmonic, the median pulses of western individuals were higher (peak frequency: 70.0 kHz versus 63.5 kHz; terminal frequency: 56.2 kHz versus 47.3

kHz). Some of the differences were a result of the difference in the emission of the fundamental.

Geographically most distant western and eastern individuals differed only in peak and terminal frequency of the second harmonic. In the second harmonic of western individuals, peak and terminal frequencies were higher than in median pulses of eastern individuals (70.0 kHz versus 64.1 kHz, 56.2 kHz versus 50.1 kHz).

The pulses of bats from geographically nearest sample sites offshore island and the central region highly differed. Except for peak frequency, all parameters differed significantly. The median pulses of central individuals had a longer duration (0.95 ms versus 0.71 ms), a lower terminal frequency (32.6 kHz versus 47.0 kHz), and a broader effective bandwidth (76.9 kHz versus 61.9 kHz). The second harmonic of median pulses of central individuals had significantly higher peak (71.3 kHz versus 63.5 kHz) and terminal (55.5 kHz versus 47.3 kHz) frequencies.

The final two comparisons between median pulses of central versus eastern, and offshore island versus eastern individuals also revealed significant differences.

Median pulses of central and eastern individuals differed in pulse duration (0.95 ms versus 0.87 ms) and in terminal frequency (32.6 kHz versus 44.1 kHz) and on parameters measured on the second harmonic; its peak and terminal frequencies were significantly lower in median pulses of eastern individuals (71.3 kHz versus 64.1 kHz; 55.5 kHz versus 50.1 kHz).

Median pulses of offshore island and eastern individuals differed significantly in peak frequency of the pulse and terminal frequency of the second harmonic. Pulses of eastern individuals had a significantly lower peak frequency (79.2 kHz versus 68.6 kHz) and significantly higher terminal frequencies of the pulse (47.0 kHz versus 44.1 kHz) and of the second harmonic (47.3 kHz versus 50.1 kHz).

In the subsequent discriminant analysis, 75% echolocation pulses from 77 individuals were correctly classified to their predetermined sample site according to their median acoustic characteristics (cross-validated method. Table 12). Misclassification of pulses from the offshore island never occurred.

Most misclassifications occurred between western and central pulses. Especially pulses of western individuals were as often assigned to the central site as to the western site. Fifteen percent of pulses of central individuals were assigned to the western site. From pulses of the 16 eastern individuals, one was assigned to west and central site each. Misclassification of eastern bats into the western or central group was explainable once more by emission of the fundamental by these two individuals, and thus, by the terminal frequency. One eastern bat which emitted the fundamental was acoustically identified as a western or central bat as terminal frequency was an acoustic characteristic of this group. Only one western individual was incorrectly classified to the eastern site.

Table 12: Classification results of the discriminant analysis. 75.3% of cross-validated individuals were correctly classified to their respective sample site.

Sample site	West	Central	Offshore island	East	Total
West	47%	47%	0%	6%	100%
	(7)	(7)	(0)	(1)	(15)
Central	15%	73%	0%	12%	100%
	(5)	(24)	(0)	(4)	(33)
Offshore island	0%	0%	100%	0%	100%
	(0)	(0)	(13)	(0)	(13)
East	6%	6%	0%	88%	100%
	(1)	(1)	(0)	(14)	(16)

The classification and grouping is reflected in Figure 12 where the individuals were plotted in the two dimensions of the first two discriminant functions according to their acoustic characteristics of echolocation pulses. Overall, the discrimination between the four groups was significant and three canonical discriminant functions separated the sites in combination (d. f. = 18, Wilk´s lambda = 0.032, p = 0.000). Although, the second and third function in combination and third function alone were also significant (Wilk´s lambda =

0.389 and 0.829, p = 0.000 and 0.010), function 1 explained 89% of the total variance (Eigenvalue = 11.016, canonical R^2 = 0.916). Function 2 explained 9 of the variance (Eigenvalue = 1.132, canonical R^2 = 0.531). In function 1,terminal frequency of the pulse scores highest (with r = 0.754), followed by terminal frequency of the second harmonic (r = - 0.316). Both terminal frequencies had a strong discriminating effect. Function 1 clearly separated offshore island population from the three others (horizontal axis in Figure 12). Function 2 (vertical axis), however, separated eastern individuals from the three other populations.

The acoustic parameters relevant for this function were terminal frequency of the second harmonic (r = 0.646), peak frequency of the second harmonic (r = 0.583) and terminal frequency of the pulse (r = 0.513). In contrast to function 1, peak and terminal frequency of the second harmonic had an opposite effect (positive values). The three remaining acoustic parameters scored moderately on function 2 (r = ≥ 0.2) in comparison to function 1 (r = ≤ 0.13).

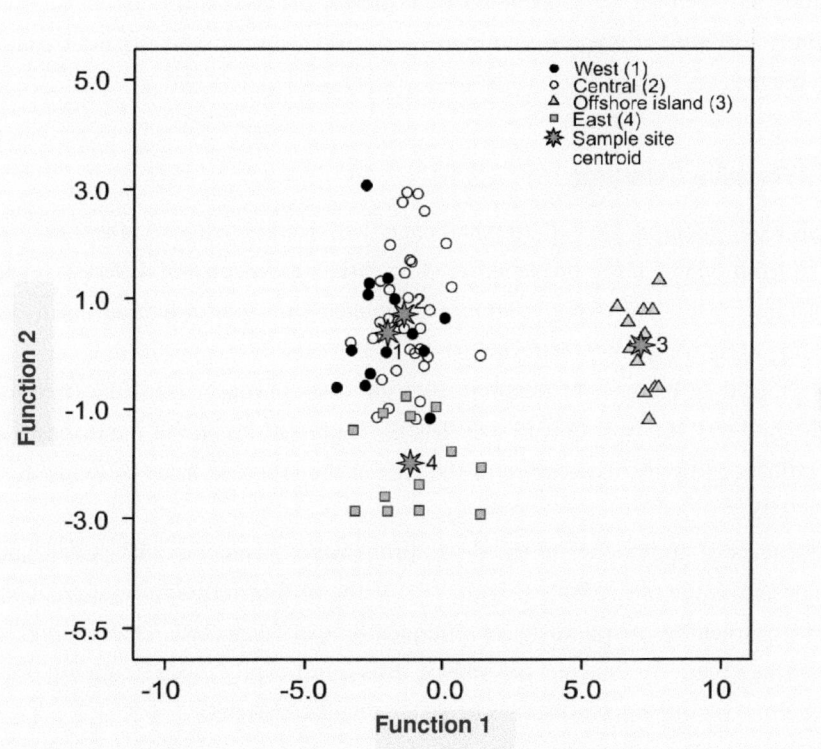

Figure 12: Plot of first two canonical functions that resulted from a discriminant analysis of multiple acoustic parameters. Measures were taken from echolocation pulses of 77 adult *M. waterhousii* from four sample sites on Cuba. Each symbol refers to one sample site.

3.3.1.1.1 Summary

To summarize these findings, I conclude that the four sites can be best differentiated acoustically by terminal frequencies and peak frequencies of the pulse and/or its second harmonic. Most different were the echolocation pulses of bats from the offshore island. Most similar were echolocation pulses from western and central bats. Echolocation pulses of bats from the east were similarly different from those emitted by central, western and offshore island individuals.

3.3.2 Passive listening

In this experiment in the experimental cage, bats could choose to take *Tenebrio* larvae from one of three presented feeding dishes each causing different levels of rustling noises. The experiment was completed with four individuals. For one bat, only eight trials were analysable.

Bats showed an increase of landings after the first approach over time (trend in Cochran-Q test, d.f. = 7, Q = 13.44). The first approach to one of the dishes was not influenced by the experience of the bats in the previous trial (Binomial test, p = > 0.05).

Frequencies of approaches to the three differently conspicuous dishes were not similar across the individuals (Figure 13). Three of four individuals approached the inconspicuous metal dish less frequently than expected. One individual approached the most conspicuous plastic dish significantly more frequently than expected. All other approaches occurred as often as expected.

The bats usually approached the dishes more than once and hovered before landing. I assumed that landing in one of the dishes required a decision of the bat which was not by chance. Total number of landings compared to number approaches (43.5% landings after approach) confirmed the assumption that the bats assure themselves to find prey before taking the cost to land. If bats landed, the percentage of successful prey capture was not significantly different between the three feeding dishes (plastic 48%, paper 42% and metal 37%). Frequencies of approaches to the three dishes without subsequent landing were not significantly different from expectations (Table 43 in the Appendix).

In two bats, a decision to land was more frequent that expected when the paper dish was approached and less frequent for the metal dish (Figure 14 Table 44 in the Appendix).

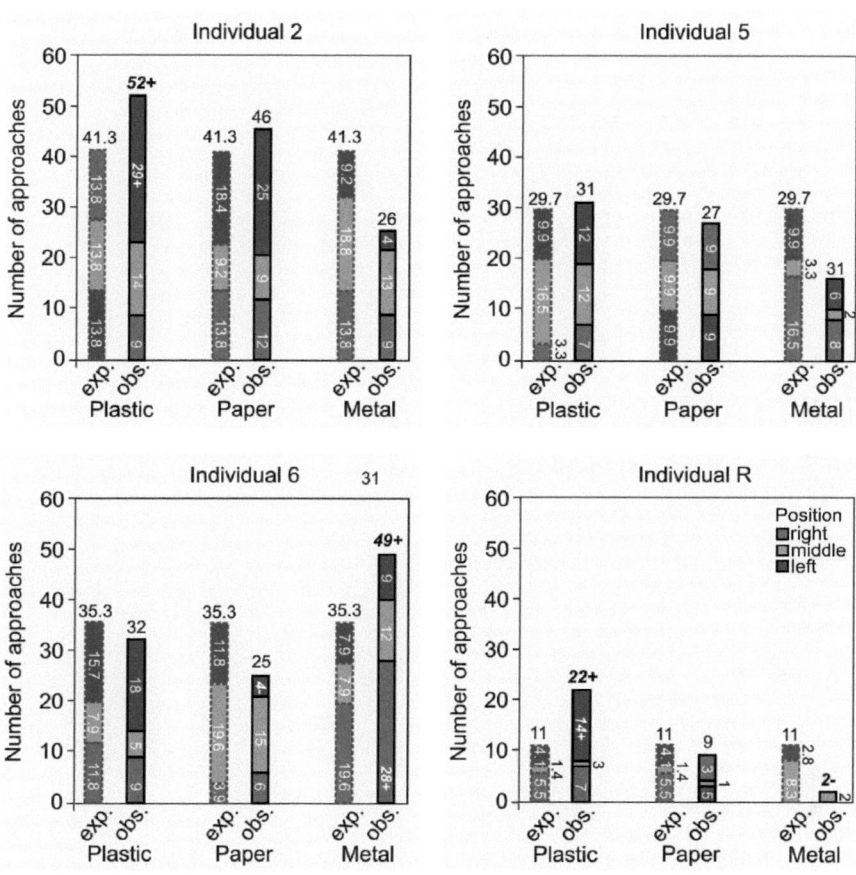

Figure 13: Number of approaches of each individual. Bar charts show expected number of approaches (exp.), assuming a random choice of feeding dishes, followed by observed number of approaches (obs.) for each feeding dish (plastic, paper, metal) on each position (left, middle, right). Bars for each material are subdivided into three parts representing the positions. Numbers above bars indicate sums of approaches for the material at the respective position. Numbers inside bars indicated expected, and observed, numbers of approaches for each material at the respective position. Significant deviations from the expected number of approaches are in bold. These observed numbers were additionally marked by an "+" when number of approaches was more frequent than expected and by an "-" when number of approaches was less frequent than expected.

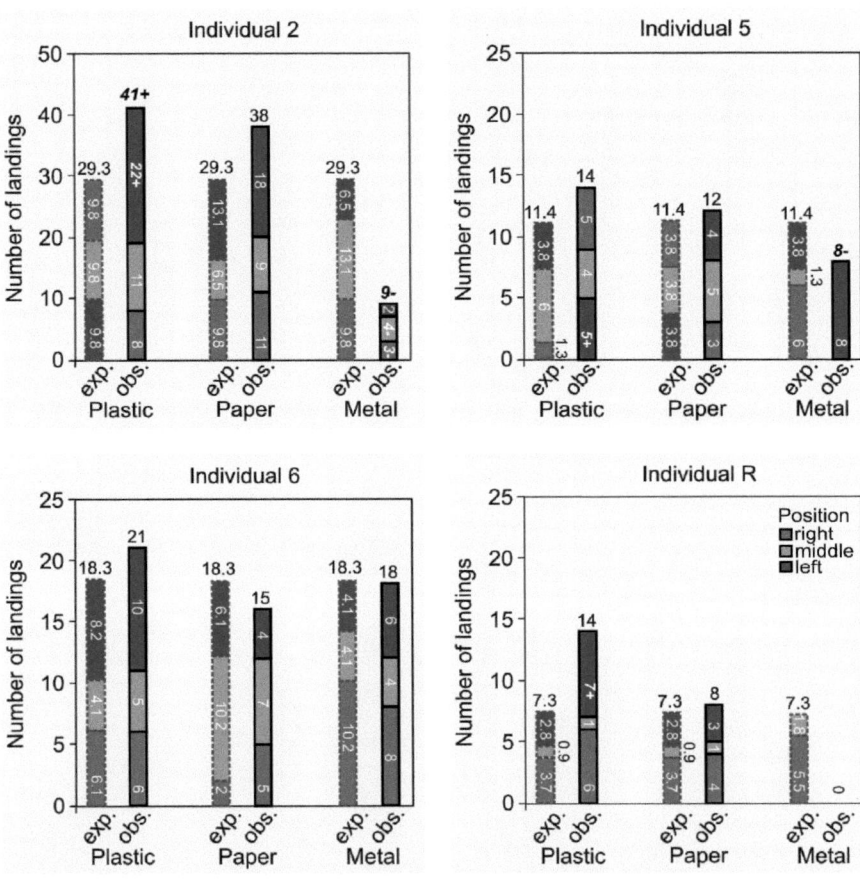

Figure 14: Number of landings for each individual. Bar charts show expected number of landings, assuming a random choice of feeding dishes, followed by observed number of landings for each feeding dish (plastic, paper, metal) on each position (left, middle, right). Bars for each material are subdivided into three parts representing the positions. Numbers above bars indicate sums of landings for the material at the respective position. Numbers inside bars indicated expected and observed numbers of landings for each material at the respective position. Significant deviations from the expected number of landings are in bold. These observed numbers were additionally marked by an "+" when number of landings was more frequent than expected and by an "-" when number of landings was less frequent than expected.

If only the first landings of each bat and each trial were analyzed, the acoustic conspicuousness of the three dishes was again decisive for the bats to land (N=33, d.f.=2, Chi²=18.2, p = < 0.05); whereas none of the four bats chose the metal dish for a first landing (N=0, d.f.=1, Chi²=11, p = < 0.05), paper was the significantly preferred dish for the first landing (N=19, d.f.=1, Chi²=5.82, p = < 0.05).

The synchronously recorded audio sequences of the approaches of bats to the feeding dishes revealed echolocation pulse emission until 57 ms before landing (based on averaged intervals of five randomly chosen recordings per bat and trial). The pulses had very low intensities (~ 10.3 db above background noise) and short durations (0.75 ms), reflecting the short distance to the target of interest.

3.3.2.1 Summary

If the bats approached the inconspicuous metal dish, they decided against a landing more often than expected. For landings, the bats chose the acoustically conspicuous plastic and paper inlayed dishes significantly more often than expected. I conclude from these results, that the Cuban *M. waterhousii* may use passive listening for finding prey. If echolocation information would have been the only source of information, the bats would have randomly chosen between the three dishes as all presented dishes contained the same amount of larvae.

All bats hovered before making the decision to land or to turn away. This hovering time allows inspecting a potential feeding source by echolocation, but more likely by listening to rustling noises, which is confirmed by preferences for the conspicuous dishes.

The role of echolocation in a foraging task without acoustic cues generated by the prey was the topic of the next experiment.

3.3.3 Prey capture success under different clutter conditions

Fourteen individuals from the three sample sites had to find dead *Tenebrio* larvae under different clutter condition. Per clutter condition, I analyzed three sessions for each of the 13 individuals at the beginning (inexperienced) and, after three days break, again three sessions (experienced), thus 18 trials per bat (and 6 trials for one individual, as there were only two analyzable session per condition). This resulted in 270 landings per bat and per condition (90 landings for one individual, respectively). For the non-clutter condition I analyzed two sessions before the moderate clutter condition, two sessions before the high clutter condition and two sessions after the high clutter condition per bat (also 18 trials per bat).

Whereas the last trial of the previous testing day did not influence the first action of the first trial on the subsequent testing day, i.e. approach to or landing on the previous reward field, subsequent trials within a session had an influence. Usually, bats inspected the previous reward field as the first action of the subsequent trial (Binomial test, $p = <0.05$, Table 45 in the Appendix). After such an unsuccessful first approach to a non-reward field, bats began to inspect the experimental field from about 30 cm. Inspection by hovering prior to a landing occurred significantly more often than by chance in most bats (Binomial tests, $p = <0.05$, Table 46 in the Appendix). Considering a learning effect on behaviour in this task, I compared number of correct landings in the reward field of first three sessions (inexperienced) and the last three sessions (experienced) for each clutter condition per bat. No significant difference was found (Friedman ANOVA, $p = \geq 0.05$). All bats were able to find dead mealworms under the three conditions and had a capture success of 87%.

In general, there was an increase in the total number of approaches and landings in the high clutter condition of all 14 bats (Table 47 and Table 48 in the Appendix). Whereas there was a significant decrease in landings and approaches between the no clutter condition versus the high clutter condition and the moderate clutter condition versus the high clutter condition in the reward field, the opposite effect was shown for landings and approaches for the non-reward fields. Thus, bats approached a non-reward field and landed more often in it with increasing clutter.

Although total number of approaches generally increased in the high clutter condition compared to the non- and moderate clutter conditions (Figure 15), this was only significant for approaches to the non-reward fields (Table 13). Approaches of the bats to the reward field even decreased.

A comparison between bats from the three sample sites revealed that sample site had no influence on the approaching behaviour of bats (Table 13).

Western, central and eastern bats slightly decreased the number of approaches to the reward field if clutter increased and increased number of approaches to the non-reward fields if clutter increased.

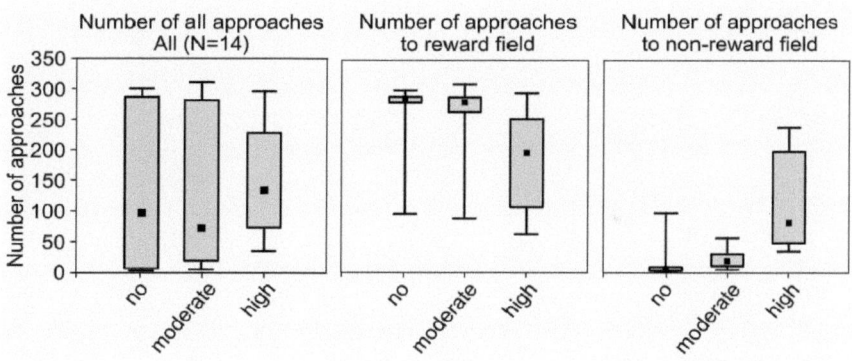

Figure 15: Boxplots showing approaches of bats to experimental fields (reward field and non-reward fields) in three different clutter conditions (no clutter, moderate clutter = small flowers, high clutter = large flowers). The upper part shows approaches of all 14 bats. The Central and the lower part show approaches of bats from the three sample sites (west, central east) separately. Figure continuned on next page.

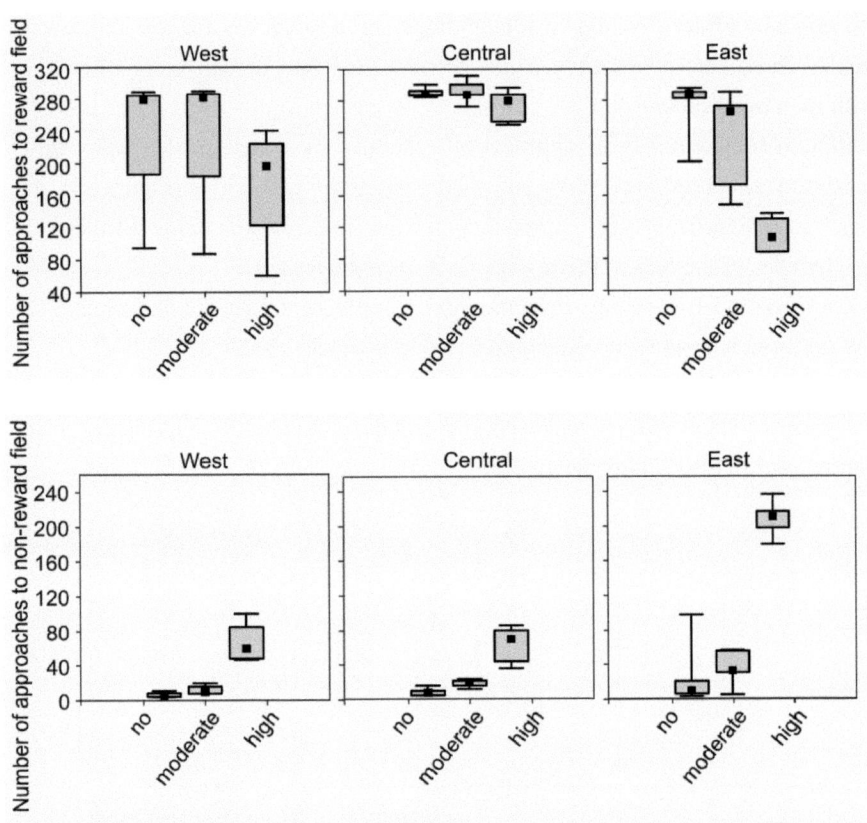

Figure 15 continued.

Table 13: Approaches of bats to the reward field or to a non-reward fields in three different clutter conditions (no clutter, moderate clutter and high clutter condition).The upper part of the table shows the result of all 14 bats comparing number of approaches in the reward and non-reward fields under the three clutter conditions using Friedman ANOVA followed by Wilcoxon matched pairs test. Statistical comparisons of approaches under the three clutter conditions within each of the three sample sites (west, central, east) follow. Due to multiple comparisons, level of significance was Bonferroni corrected. Table continued on next page.

All reward

Friedman ANOVA Chi² (N = 14. d.f. = 2) = 13.0. p = 0.002	Average rank	Mean	Std.Dev.
no clutter reward	2.4286	268.79	55.023
moderate clutter reward	2.3571	254.07	66.776
high clutter reward	1.2143	187.36	83.059
Wilcoxon matched pairs test			n.s.

All non-reward

Friedman ANOVA Chi² (N = 14. d.f. = 2) = 26.14 p = 0.000	Average rank	Mean	Std.Dev.
no clutter non-reward	1.0714	14.29	24.910
moderate clutter non-reward	1.9286	22.50	16.709
high clutter non-reward	3.0000	116.21	75.809
Wilcoxon matched pairs test	T	Z	p
no clutter non-reward & moderate clutter non-reward			n.s.
no clutter non-reward & high clutter non-reward	0	3.296	0.001
moderate clutter non-reward & high clutter non-reward	0	3.296	0.001

West reward

Friedman ANOVA	n.s.

West non-reward

Friedman ANOVA	n.s.

Central reward

Friedman ANOVA	n.s.

Central non-reward

Friedman ANOVA Chi² (N = 5. d.f. = 2) = 10.0. p = 0.007	Average rank	Mean	Std.Dev.

Table 13 continued.

Wilcoxon matched pairs test	n.s.
East reward	
Friedman ANOVA	n.s.
East non-reward	
Friedman ANOVA	n.s.

The landing behaviour of all bats was similar to the approaching behaviour; there was a general decrease of landings in the reward field in the high clutter condition and an increase of landings in the non-reward fields (Figure 16).

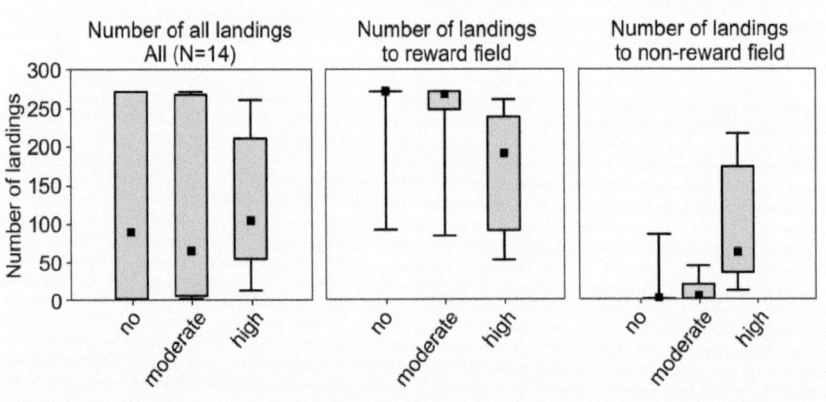

Figure 16: Boxplots showing landings of bats to experimental fields (reward field and non-reward fields) under three different clutter conditions (no clutter, moderate clutter = small flowers, high clutter = large flowers). The upper part shows landings of all 14 bats. The central and the lower part show landings of bats from the three sample sites (west, central, east), separately. Figure continued on next page.

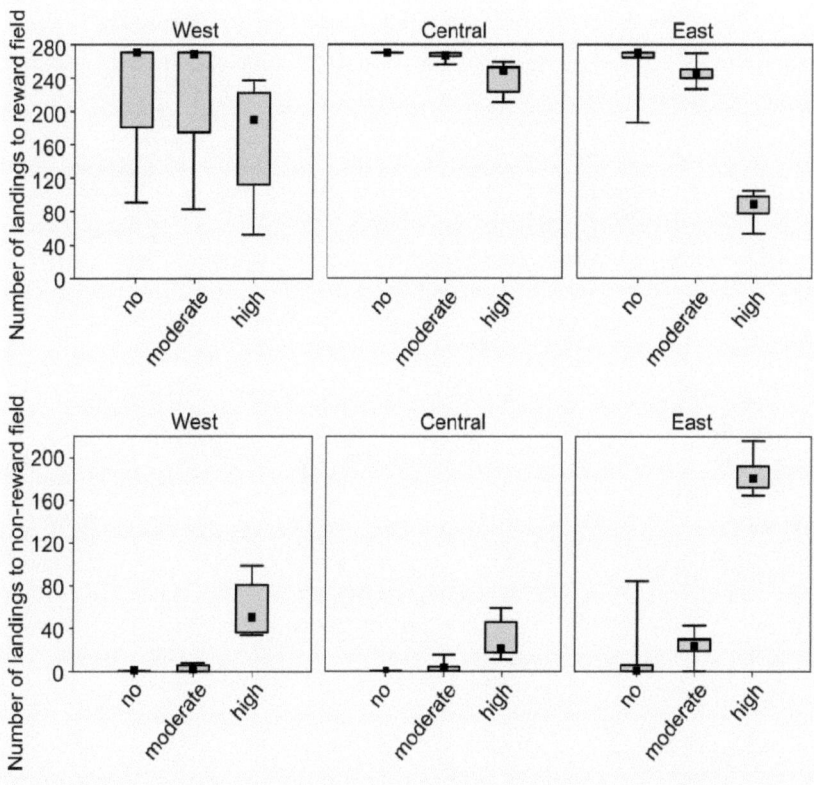

Figure 16 continued.

A comparison of clutter conditions within bats from one sample site revealed that these differences were significant for central and for eastern bats, but not for western bats (Table 14). In general, bats from the western sample site had not many landings in a non-reward field, and not more in the moderate clutter condition compared to the non-clutter condition. Although the general increase in approaches to the non-reward field in the high clutter condition applied also for the western sample, differences were not significant. Within the sample of the central individuals, approaches and landings decreased slightly for the reward field in the high clutter condition. Landings in non-reward fields occurred more often in the high clutter condition, but this result was not significant.

Table 14: Landings of bats in the reward field or in the non-reward fields in three different clutter conditions (no clutter, moderate clutter and high clutter condition). The upper part of the table shows the result of all 14 bats. Results for each individuals and every samples site follow. Number of landings in the reward and non-reward fields under the three clutter conditions were compared using Friedman ANOVA followed by Wilcoxon matched pairs test (Bonferroni corrected). Table continued on next page.

All reward			
Friedman ANOVA Chi² (N = 14, d.f. = 2) = 24.98 p < 0.000	Average rank	Mean	Std.Dev.
no clutter reward	2.7500	250.71	51.355
moderate clutter reward	2.2500	246.50	49.252
high clutter reward	1.0000	163.21	79.798
Wilcoxon matched pairs test	T	Z	p
no clutter reward & high clutter reward	0	3.296	0.001
moderate clutter reward & high clutter reward	0	3.296	0.001
All non-reward			
Friedman ANOVA Chi² (N = 14, d.f. = 2) = 26.14 p = 0.000	Average rank	Mean	Std.Dev.
no clutter non-reward	1.2500	6.429	22.384
moderate clutter non-reward	1.7500	10.714	13.612
high clutter non-reward	3.0000	93.929	74.663
Wilcoxon matched pairs test	T	Z	p
no clutter non-reward & high clutter non-reward	0	3.296	0.001
moderate clutter non-reward & high clutter non-reward	0	3.296	0.001
West reward			
Friedman ANOVA			n.s.
West non-reward			
Friedman ANOVA			n.s.

Table 14 continued.

Central reward

	Average rank	Mean	Std.Dev.
Friedman ANOVA Chi² (N = 5, d.f. = 2) = 9.33 p < 0.0094			
no clutter non-reward	2.8000	270.00	
moderate clutter non-reward	2.2000	265.80	5.762
high clutter non-reward	1.0000	239.20	20.645
Wilcoxon matched pairs test			n.s.

Central non-reward

	Average rank	Mean	Std.Dev.
Friedman ANOVA Chi²			
no clutter non-reward & moderate clutter non-reward	1.2000	0.000	
no clutter non-reward & high clutter non-reward	1.8000	4.400	6.189
moderate clutter non-reward & high clutter non-reward	3.0000	30.800	20.645
Wilcoxon matched pairs test			n.s.

East reward

Friedman ANOVA (N = 5, d.f. = 2) = 8.32, p < 0.0156

no clutter non-reward & moderate clutter non-reward	2.700000	252.0000	36.98648
no clutter non-reward & high clutter non-reward	2.300000	246.8000	15.77023
moderate clutter non-reward & high clutter non-reward	1.000000	84.8000	19.96747
Wilcoxon matched pairs test			n.s.

East non-reward

Friedman ANOVA (N = 5, d.f. = 2) = 8.32 p < 0.0156

no clutter non-reward & moderate clutter non-reward	1.300000	18.0000	36.98648
no clutter non-reward & high clutter non-reward	1.700000	23.2000	15.77023
moderate clutter non-reward & high clutter non-reward	3.000000	185.2000	19.96747
Wilcoxon matched pairs test			n.s.

3.3.3.1 Summary

The increase of acoustical challenge to find dead prey by echolocation within clutter had an effect on approaches to, and landings of the bats on reward fields. While most bats even decreased their approaches and landings in the reward field, they increased their approaches and landings to the non-reward fields. Approaching and landing in the non-reward fields indicated that bats had more difficulties to find dead mealworms in the presence of the highly cluttered background. Furthermore, there was a difference between bats from the three sample sites: The different clutter conditions showed no significant effect on western individuals. Their prey capture performance remained similar between the three clutter conditions. Central bats changed their behaviour with increasing clutter according to the described general effect. These bats had more difficulties to find prey in the high clutter condition. Individuals from the eastern group had most difficulties to find and capture prey in the high clutter condition as they approached, and often landed in, the non-reward fields.

An influence of body size on the approaches and/or landing behaviour was not likely, as all bats folded their wings before landing. All bats landed frequently on the testing area and showed no differences in the uplift from the ground.

3.3.4 Summary of acoustic orientation and foraging behaviour

In a standardized situation, the acoustic structure of echolocation pulses from individuals from the four sample sites differed. Acoustically similar were individuals from the west and the central area. Both groups had only minor difficulties to capture prey in a high clutter condition. Eastern bats differed from western/central bats in peak frequency of the second harmonic and terminal frequency of the pulse and of the second harmonic. Furthermore, eastern bats differed acoustically more from central individuals than from western individuals. Eastern bats had a 14-kHz-narrower effective bandwidth due to higher terminal frequencies and were less successful to find and capture prey in a highly cluttered environment.

3.4 Social interactions

3.4.1 Spatial allocation

The positions of each bat inside the flight room was analysed from 400 snap shots.
I tested the influence of time across the four recording months on the spatial allocation of bats and found no significant effect (Friedman ANOVA (d.f.= 7, p = > 0.05).
In general, bats did not use all areas in the facility as perches (Figure 17). The feeding area and the roost were less frequently used as a perch than a random perch usage of all possible areas would predict. Vice versa, there were preferred areas by the bats, namely cable 1, cable 3 and the edge of the wall. Bats used the wall, cable 2, and the ceiling as often as expected.
The individuals had different preferred areas in the flight room (see Table 49 in the Appendix).

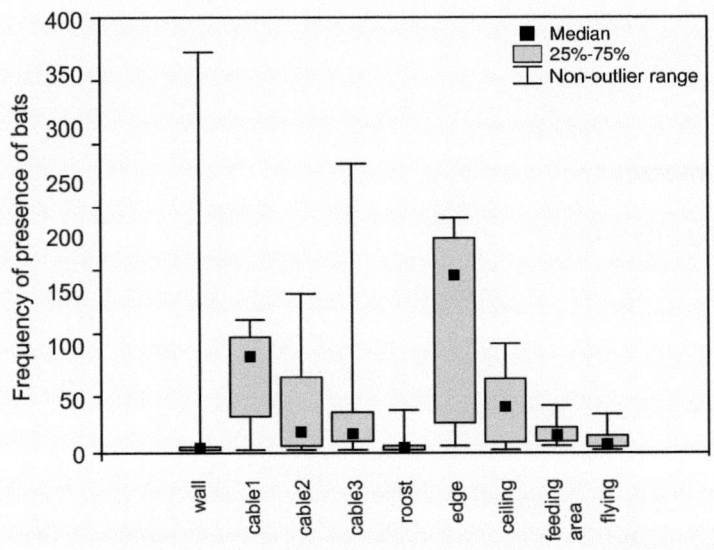

Figure 17: Presence of bats at defined areas inside the flight room. Box plots show median, lower and upper quartiles and non-outlier ranges of frequencies of bats, present in each area.

On the level of sample site, individuals from the same sample site showed preferences for a specific area (Table 15). All western bats preferred the wall, two in addition cable 1, one the edge of the wall and one preferred cable 2. Central bats preferred cable 1 and cable 2. All bats from the east preferred cable 1 and the edge of the wall. Six bats also preferred the ceiling. In addition, I found that the preferred areas of one group were rarely-used areas of the other two groups. Thus, central and western bats shared cable 2, whereas western and eastern bats shared cable 1. Individuals caught at same sample site clustered together in a specific area, although areas were not exclusively occupied by one sample site group and preference for some areas overlapped between groups.

Table 15: Observed frequencies and respective test values for each predefined area in the housing facility separated for west, central and eastern bats.

wall		cable1		cable2		cable3		roost		edge		ceiling		feeding area		flying	
Obs.	Exp.	Obs.	Exp.	Obs.	Exp.	Obs.	Exp.	Obs.	Exp.	Obs.	Exp.	Obs.	Exp.	Obs.	Exp.	Obs.	Exp.
West																	
78	25	95	58	45	0.01	15	20	9	28	91	49	40	0.44	14	21	13	22
370	238	0	44	0	44	0	44	0	44	4	37	0	44	21	12	5	35
231	783	15	20	17	17	0	44	5	35	97	62	31	4.06	4	37	0	44
182	426	105	83	88	43	0	44	0	44	5	35	7	32	9	28	4	37
	362		204		104		153		152		183		81		98		139
Central																	
0	44	57	3.50	101	72	100	70	37	1.25	31	4.06	0	44	41	0.27	33	2.94
0	44	31	4.06	145	228	167	338	0	44	11	25	2	41	31	4.06	13	22
0	44	22	11	97	62	214	647	3	39	25	9	6	33	14	21	19	15
1	43	31	4.06	88	43	237	834	17	17	8	30	5	35	8	30	5	35
0	44	13	22	68	13	266	1105	12	24	11	25	8	30	21	12	1	43
	220		46		417		2993		125		93		183		67		117
East																	
0	44	88	43	3	39	31	4.06	0	44	211	624	57	3.55	7	32	3	39
2	41	97	62	1	43	3	39	2	41	204	573	66	11	21	12	4	37
3	39	107	88	6	33	15	20	2	41	182	426	78	25	6	33	1	43
0	44	111	100	14	21	17	17	1	43	209	609	39	0.67	9	28	0	44
2	41	121	132	13	22	13	22	0	44	162	311	74	20	11	25	4	37
9	28	95	58	8	30	14	21	0	44	157	285	99	67	16	18	2	41
0	44	86	39	0	44	7	32	2	41	197	524	89	45	10	27	9	28
0	44	109	94	4	37	12	24	4	37	191	48	55	2.51	8	30	17	17
3	39	116	115	21	12	35	2.01	0	44	166	333	37	1.25	7	32	15	20
0	44	79	27	23	10	18	16	2	41	199	538	55	2.51	17	17	7	32
2	41	87	41	0	44	8	30	1	43	216	662	55	2.51	21	12	10	27
2	41	75	21	22	11	21	12	0	44	179	407	69	14	19	15	13	22
	490		819		347		23		506		5775		194		281		385

3.4.2 Allogrooming partners

Allogrooming occurred frequently in the social group. In analysed 40 hours, I recorded 89 allogrooming events, thus it occurred on average twice per hour. Interaction partners were identifiable in 72 allogrooming events. Fifteen of 21 individuals participated in allogrooming events (Table 50 in the Appendix). One western individual, two central and three eastern individuals were never observed allogrooming. Typically both interaction partners actively participated in allogrooming. Six bats allogroomed ≥5 times and four bats ≥2 times. Most frequently one western individual ("A") and one eastern individual ("S") were observed in allogrooming events (14 respectively 10 times). Of 98 possible dyads between the 14 individuals, 44 different dyads occurred (Table 16) resulting in an overall dyad diversity of 45%.

Table 16: Summarized results of allogrooming dyads within a group of the same sample size or between different sample site members. The allogrooming coefficient reflects allogrooming activity for the respective group and thus allows comparisons.

Number of allogrooming events	Basis of calculation	Allogrooming coefficient (events / n of group)	Dyad diversity (observed / possible number of dyads)
19	West (n=3)	6.33	49%
9	West with west	3.00	100%
4	West with central	1.33	44%
6	West with east	0.67	22%
16	Central (n=3)	5.33	41%
9	Central with central	3.00	100%
4	Central with west	1.33	44%
7	Central with east	0.78	26%
37	East (n=9)	2.64	41%
22	East with east	2.44	61%
6	East with west	2.00	22%
7	East with central	2.33	26%

All participating bats allogroomed with individuals from different sample sites, but the percentage of allogrooming with sample site members was higher. Within the group of western bats, allogrooming occurred to 100%. Allogrooming with central bats occurred to 44% and to 22% with eastern bats. Within the group of central bats, allogrooming occurred also to 100% and to 26% with eastern bats. Within the group of eastern bats, allogrooming occurred to 61%. Percentage of bats participating in all possible allogrooming events was similar across the three groups, 49% for western bats, 41% for central and eastern bats. Statistical testing was not feasible due to zero frequencies for several possible dyads. The influence of time across the four months on allogrooming dyads revealed that allogrooming events of mixed-site dyads increased from 23% (17 events) to 34% (25 events).

The low overall dyad diversity might be the result of individually preferred allogrooming partners or in addition, preference for individuals from the same sample site. The preference for sample site members diminished over time.

3.4.3 Social calls

Social calls were emitted occasionally from all individuals in the flight room and of individuals inside the transportation cage. I obtained a data set of 37 social interactions with accompanying social calls of acceptable quality. Nine calls were emitted by the western bats, 13 by central bats and 15 by eastern bats.

The bats emitted three call types (). One common call type consisted of frequency downward modulated syllables uttered in a series of variable length. An average syllable of this call type was 4.2 ms of duration, multiharmonic, with peak frequency at the fundamental of 34.4 kHz or at the second harmonic of 65.2 kHz. A second call type consisted of U-shaped syllables which were uttered in bouts and bouts could be uttered in series. A u-shaped syllable had a duration of 7.3 ms, was multiharmonic with peak frequencies of 25.4 kHz at the fundamental or 48.2 kHz at the second harmonic.

The third frequently emitted call type consisted of single syllables. It was a multiharmonic syllable with a long duration of 84.7 ms and was often repeatedly emitted. It had an irregular sinusoidal time frequency contour and a peak frequency of 22.8 kHz at the fundamental harmonic. U-shaped syllables and sinusoidal syllables might be seen as a continuum as they often merged into another.

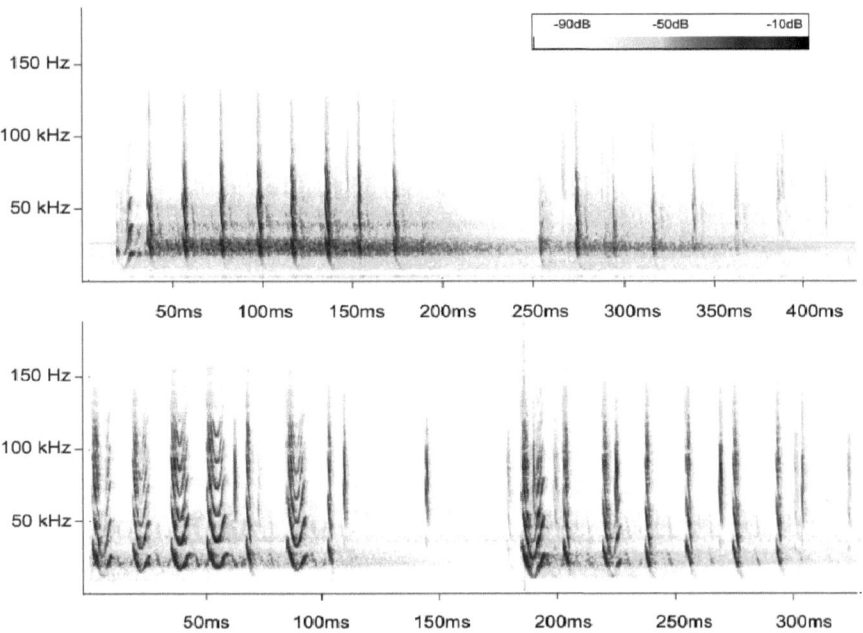

Figure 18: Sonagrams of typical calls of three commonly emitted social call types. The upper sonogram shows a call type that consists of frequency downward modulated syllables. Please not that first syllable is of a u-shaped type. The central sonagram shows a social call type that consists of u-shaped syllables in combination of fm-downward modulated syllables at the end of the call. The last sonagram shows a call type that consists of sinusoidal syllables. Figure continued on next page.

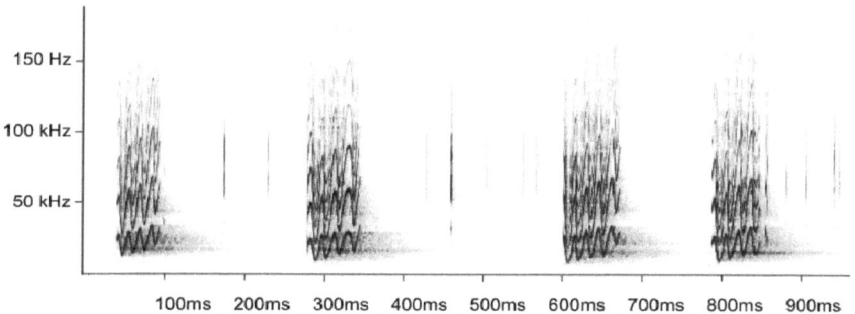

Figure 18 continued

The three described call types, respectively syllable types occurred in different contexts of presumably high arousal states of the calling bat. Those contexts were a) approaches to another bat which was at a perch or on the wing, b) after the experiments when the group was reunited I often observed a single bat landing isolated at a perch emitting the sinusoidal syllables, c) or if a group of bats were kept in transportation cages for more than two hours. Except for b), all of the syllable/call types occurred in the mentioned contexts with equal frequencies.

The syllable/call types did not show sample site specific acoustic signatures (Kruskal Wallis test, $p = <0.05$). An analysis of individual signatures was not feasible as number of high quality recordings was too low.

3.4.4 Summary

Bats from the same sample site preferred the same areas in the facility and allogroomed more often than with bats from other sample sites. Group affiliation was given, but the effect of sample site became weaker over time, thus individuals from different sample sites mixed more often. Within the three social call types, acoustic parameters did not show a difference between individuals from the sample sites.

3.5 Genetic variability

The 20 sequences in the cyt-b alignment were unequivocal, had 420 characters with two gaps. 276 characters were constant, 100 were parsimony-informative (for an overview of Indels (*in*sertions and *del*etions), see additional descriptive results of sequence analyses Table 51 in the Appendix). Sequence differences led to seven different haplotypes (overall $h = 0.8474$) based on single nucleotide polymorphisms (Table 17). Number of haplotypes per sample site ranged from one to three and no haplotype was shared between the sites (Table 52 in the Appendix).

Table 17: Summary statistics of Cytochrom-b sequence data of Cuban M. waterhousii. The upper part gives descriptive statistics. The part below gives measures of sequence variation.

Cyt-b 418 bp fragment	total	west	central	island	east
N = number of sequences	20	5	5	5	5
S = variable sites	19	0	2	2	0
$S_\%$ = % of variable sites	4.5	0	0.5	0.5	0
TI = number of transitions	19	0	2	2	0
TV = number of transversions	0	0	0	0	0
π **= nucleotide diversity**	0.016281 +/-0.008940	0 +/- 0	0.002392 +/- 0.002227	0.001914 +/- 0.001906	0 +/- 0
A = number of haplotypes	7	1	3	2	1
h **= haplotype diversity**	0.8474 +/- 0.0421	0 +/- 0	0.7000 +/- 0.2184	0.4000 +/- 0.2373	0 +/- 0
Label and frequency of haplotypes	-	W1, 5	C1, 1; C2,1; C3, 3	I1,4; I2,1	E1,5
Tajima´s D and P	1.02374; 0.88100	0; P=1	0.24314; P=0.72200	-0.97256; P=0.18600	0; P=1
Pi = mean number of pairwise differences	6.805 +/-0	0	1.0	0.8	0
Coefficient of differentiation	0.934 +/- 0.034				

In the west, all individuals had the same haplotype as well as eastern bats. Within the single cave on the offshore island two haplotypes occurred. The highest number of haplotypes occurred in the central region with three haplotypes.

A minimum spanning network of the 20 individuals revealed that island haplotypes are most distant to the central sample site, the geographically closest, but only two, respectively four, steps were needed to the eastern haplotype (Figure 19). Haplotypes of the central region are most distant to the others, requiring nine mutational steps to the western haplotype.

The topology of the phylograms of the Neighbor Joining, Maximum Parsimony and Maximum Likelihood analyses was identical (Figure 20). Bootstrap analyses revealed well supported clades of the designated outgroups (100% bootstrap support), and for the clade consisting of individuals from central Cuba (>90%). The node referring to the branch comprising western, offshore island and eastern individuals was less strongly supported (67% and 64%, respectively), the next level (e.g. sub-) clade of individuals from the west was well supported (> 90%). The subclade of eastern individuals was, again, well supported (> 80%), but individuals from the offshore islands were poorly resolved.

The overview of sequence variation based on the straightforward uncorrected p-distances is shown in Table 18 summarizing percentages of differences within sample sites, between the sample sites, among Cuban *M. waterhousii*, between Cuban *M. waterhousii* and the outgroups.

Table 18. Estimates of evolutionary divergence between cyt-b sequences. The number of base differences per site from analysis between sequences is shown ("uncorrected p"). All results are based on the pairwise analysis of 25 sequences. Analyses were conducted in MEGA4. Codon positions included were 1st+2nd+3rd+Noncoding. All positions containing gaps and missing data were eliminated from the dataset (Complete deletion option). There were a total of 418 positions in the final dataset. Given are estimates of Average Evolutionary Divergence over Sequence Pairs within Groups. The number of base differences per site from averaging over all sequence pairs within each group is shown. Standard error estimate(s) are shown in the second column and were obtained by a bootstrap procedure (1000 replicates).

Comparison basis	P-distance	Remarks
within west	0.0000 +/-0.0000	
within central	0.0024 +/-0.0017	
within island	0.0019 +/-0.0013	
within east	0.0000 +/-0.0000	
within Cuba mean	0.0163 +/-0.0037	
between c-e	0.0325 +/-0.0079	one island
between c-i	0.0287 +/-0.0075	one island
between w-c	0.0254 +/-0.0071	one island
between w-e	0.0167 +/-0.0060	one island
between w-i	0.0129 +/-0.0053	one island
between i-e	0.0058 +/-0.0031	one island
between Cuban sites mean	0.0152 +/-0.0037	one island
between cuba-mex	0.1221 +/-0.0151	sub?-sister?-species
between cuba-calif	0.1829 +/-0.0177	sister species
between cuba-micronyc	0.18468 +/-0.0186	genus
between cuba-uroder	0.1981 +/-0.0186	genus
between cuba-phyll	0.2003 +/-0.0186	genus
between mex-calif	0.1603 +/-0.0177	sister species

The average sequence diversity was 0.016 +/-0.004 among the 20 Cuban *M. waterhousii* sequences. This value decreased to 0.001 +/- 0.001 if calculation was based on sequence diversity within the four sample sites. Accordingly, a comparison of sequence diversity between the sample sites resulted in an increased value of 0.015 +/-0.004. The average sequence divergence within Cuban sample sites was almost half the amount of average between Cuban sample site divergence (0.0163 to 0.0203). I calculated a coefficient of differentiation of 0.934 +/-0.036 for cyt-b sequences of Cuban *M. waterhousii*. In comparison to the most closely related outgroup, *M. waterhousii* from Mexico, sequence divergence based on p-distances is 6-fold (between Cuban sites 0.0203 and 0.1221 between Cuban and Mexican *M. waterhousii*). Amount of sequence divergence between Cuban *M. waterhousii* and the different outgroups increased according to increasing taxonomic relatedness (*M. californicus*: 0.1829; *Micronycteris megalotis*: 0.1846; *Uroderma magnirostrum*: 0.1981; *Phyllostomus hastatus*: 0.2003). Although genetic differences between the four sample sites were small, pairwise comparisons of genetic differences, FST values, and respective significance statistics revealed differences between the four sampling sites (Table 19).

Furthermore, AMOVA Table 20) confirmed that 95% of the molecular variance in the data set is due to variation among the samples sites whereas only 5% was due to variation within the sample sites. This relatively high degree of genetic structuring among Cuba (overall FST = 0.947) was weakly related to geographic distances (Mantel test critical value 1.645, $g = 0.65$, $Z = 877.26$, $r = -0.29$, 999 permutations).

Table 19: Between sample site genetic pairwise differences of cyt-b sequences. P values were calculated by using 110 permutations and were corrected for multiple comparisons by Holme's sequential Bonferroni correction = 0.008; 0.01; 0.0125; 0.0167; 0.025; 0.05.

pairwise F_{ST}'s (upper part) and P values (lower part)	west	central	island	east
west		0.95283	0.92593	1.00000
central	0.01802*		0.92500	0.96324
island	0.00901*	0.02703*		0.83333
east	0.00000*	0.00901*	0.00901*	

Table 20: Results of AMOVA. F_{ST} value is based on haplotype frequencies of cyt-b sequences.

AMOVA	d.f.	Sum of squares	Variance components	% of total variance
Among populations	3	61.050	4.025	94.71
Within populations	16	3.600	0.225	5.29
Total	19	64.650	4.250	100
Fixation index F_{ST} and respective P value (1023 permutations)	0.947; 0.000			

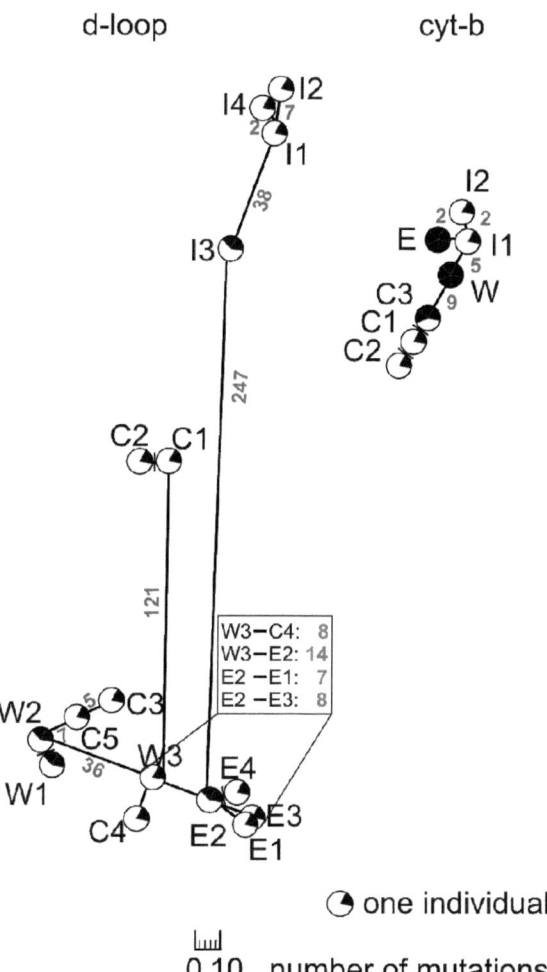

Figure 19: Minimum spanning haplotype network of the d-loop region (on the left) and cyt-b gene (on the right) of sequences from 20 Cuban M. waterhousii. Each haplotype is represented by a circle. One black filled fifths of a circle refers to one individual. Names of haplotypes start with the letter of the sample site (W = west, C = central, I = offshore island and E = east) followed by a number. Numbers in grey are numbers of mutational steps which are needed to connect two haplotypes.

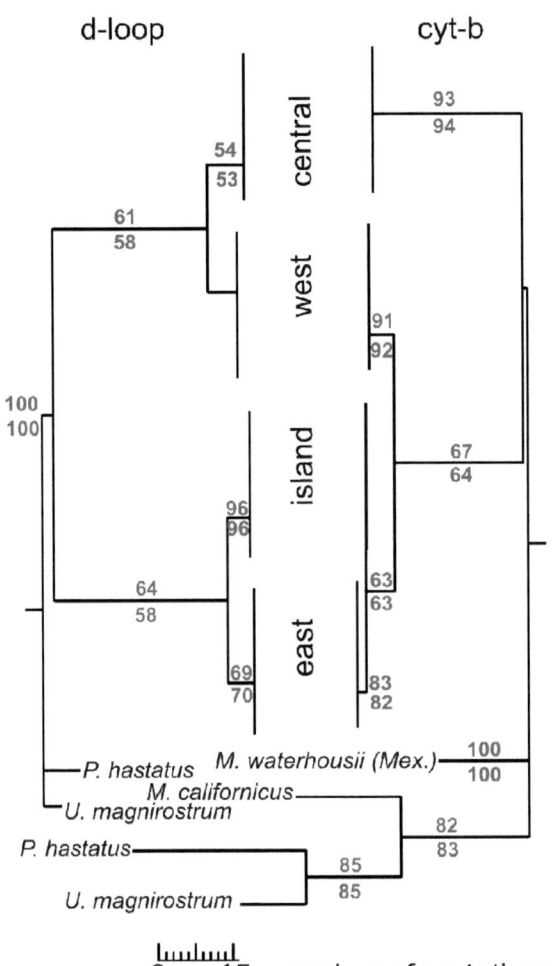

Figure 20: Phylogram of d-loop and cyt-b of sequences from 20 Cuban M. waterhousii from four different sample sites and designated outgroups. Tree topology was consistent throughout NJ, MP and ML analyses. Shown here is the NJ tree. Numbers above the branches give bootstrap support (in %) obtained within MP and ML methods. Length of a branch (y-axis) denotes the genetic distance, e.g. number of substitutions between two taxa it connects.

Multisequence alignment of a 400 bp fragment of the d-loop region had 211 constant, 103 parsimony-informative characters and three gaps. Of all substitutions, transversions outnumbered transitions (225 to 139. For an overview of indel events, see Table 53).

The diversity of nucleotides ($\pi = 0.327$) in the data set resulted in 16 haplotypes (overall $h = 0.9789$). Like in cyt-b, none of the haplotypes was shared between the four sample sites, but in contrast to cyt-b, almost each individual had a distinct haplotype (Table 21). This high diversity makes it likely that a NUMT (*nu*clear *mit*ochondrial DNA) was amplified. As there is no published d-loop sequence for Macrotus available, this assumption cannot be verified.

Table 21: Summary statistics of d-loop sequence data of Cuban M. waterhousii. The upper part gives descriptive statistics. The part below gives measures of sequence variation. Table continued on next page.

D-loop 379 bp fragment	total	west	central	island	east
N = number of sequences	20	5	5	5	5
S = polymorphic sites	42	1	8	11	14
TI = number of transitions	139	16	40	26	13
TV = number of transversions	225	21	123	21	1
π = nucleotide diversity	0.32699 +/-0.16361	0.039894 +/- 0.025231	0.255319 +/- 0.155698	0.070213 +/- 0.043600	0.015957 +/- 0.010705
A = number of haplotypes	16	3	5	4	4
h = haplotype diversity	0.9789 +/- 0.0214	0.8000 +/- 0.1640	1.0000 +/- 0.1265	0.9000 +/- 0.1610	0.9000 +/- 0.1610
Label and frequency of haplotypes	-	W1,2; W2,2; W3,1	C1,1; C2,1; C3,1; C4,1; C5,1	I1,1; I2,1; I3,2; I4,1	E1,1; E2,2; E3,1; E4,1
Tajima´s D and P	2.25481; P = 0.97900	-1.16701; P = 0.06600	1.90501; P = 0.99700	1.28333; P = 0.89900	-0.78089; P = 0.28900

Table 21 continued.

Pi	122.94737	15.00000	96.00000	26.40000	6.00000
Coefficient of differentiation	0.727 +/- 0.040				

Lowest number of haplotypes occurred in the west with three haplotypes. In the eastern and the island sample four haplotypes occurred at each site. In the central sample each individual was represented by a distinct haplotype.

The minimum spanning network revealed that haplotypes of island individuals were most distant to the others by > 200 mutations (Figure 19) the pattern between haplotypes of western, central and eastern haplotypes was less clear than for cyt-b. Haplotypes from the offshore island were most distant from the others. Closest connection was to eastern haplotype (E4) with more than 200 mutations. However, apart from the offshore island sample, number of mutational steps between haplotypes from the same sample site sometimes exceeded the number of mutational steps needed to connect haplotypes from different sample sites. Whereas two haplotypes of individuals from the central region are 121 mutational steps from the closest western haplotype, differences of remaining central haplotypes to closest western haplotypes were not greater than to next central haplotypes (max. 9 and max. 36 mutations). Eastern haplotypes never required more than eight mutations between them, but 14 to the closest western haplotype.

Consistent tree topologies of Maximum Parsimony and Maximum Likelihood analyses revealed three major clades (Figure 20); Highest support was found for the clade consisting of individuals from the offshore island (>90%). The clade formed by individuals from the east achieved also good support values (>69%) and the clade comprising individuals from the west and the central region was supported by 58-60%. The NJ method resulted in a tree topology of d-loop region which differed on the first level of clades from the cyt-b tree; island and eastern individuals formed one clade subdivided into separate clades in the next step and western and central individuals formed a second clade, further subdivided into two subclades.

The overview of sequence variation based on the straightforward uncorrected p-distances is shown in Table 22, summarizing percentages of differences within sample sites, between the sample sites, among Cuban *M. waterhousii*, between Cuban *M. waterhousii* and the outgroups.

Table 22. Estimates of evolutionary divergence between d-loop sequences. The number of base differences per site from analysis between sequences is shown ("uncorrected p"). All results are based on the pairwise analysis of 25 sequences. Analyses were conducted in MEGA4. Codon positions included were 1st+2nd+3rd+Noncoding. All positions containing gaps and missing data were eliminated from the dataset (Complete deletion option). There were a total of 418 positions in the final dataset. Estimates of Average Evolutionary Divergence over Sequence Pairs within Groups. The number of base differences per site from averaging over all sequence pairs within each group is shown. Standard error estimate(s) are shown in the second column and were obtained by a bootstrap procedure (1000 replicates).

Comparison basis	P-distance	Remarks
within west	0.002 +/-0.002	
within central	0.010 +/-0.003	
within island	0.013 +/-0.004	
within east	0.016 +/-0.004	
within Cuba mean	0.010 +_/- 0.002	
between c-i	0.055 +/-1.010	one island
between i-e	0.055 +/-0.010	one island
between c-e	0.047 +/-0.009	one island
between w-e	0.043 +/-0.010	one island
between w-i	0.042 +/-0.009	one island
between w-c	0.021 +/-0.006	one island
between Cuban sites mean	0.027 +/-0.005	one island
between cuba-uroder	0.258 +/-0.022	genus
cuba-phyll	0.277 +/-0.022	genus

Although genetic differences between the four sample sites were again small, compared to distances between Cuban specimens and outgroups, pairwise comparisons of genetic differences, FST values, and respective significance statistics revealed differences between the four sampling sites like in the cyt-b analysis (Table 23).

Table 23: Between sample site genetic pairwise differences of d-loop sequences. P values were calculated by using 110 permutations and were corrected for multiple comparisons by Holme´s sequential Bonferroni correction = 0.008; 0.01; 0.0125; 0.0167; 0.025; 0.05.

pairwise FST·s (upper part) and P values (lower part)	west	central	island	east
west		0.24592	0.90704	0.75236
central	0.07207		0.72814	0.32253
island	0.00000	0.00000		0.93192
east	0.00901	0.00901	0.00000	

Furthermore, AMOVA (Table 24) confirmed that 75% of the molecular variance in the data set is due to variation among the samples sites whereas 25% was due to variation within the sample sites. This genetic structuring among Cuba (overall FST = 0.75) was not related to geographic distances (Mantel test critical value 1.645, g = -0.0892, Z = 314.0265, r = -0.0446, 999 permutations).

Table 24: Results of AMOVA. F_{ST} value is based on haplotype frequencies of d-loop3 sequences. P values (110 permutations) significance levels Holme´s sequential Bonferroni = 0.008; 0.01.

AMOVA	d.f.	Sum of squares	Variance components	% of total variance
Among sites	3	881.200	55.161667	75.47
Within sites	16	286.800	17.92500	24.53
Total	19	1168.000	73.08667	
Fixation index F_{ST}	0.75474; P = 0.00000			

The congruence of topologies of phylogenetic consensus trees between the two genetic markers was not complete. Both trees had in common that sequences of individuals from the same sample site formed one terminal clade each, and the split of Cuban *M. waterhousii* from the outgroups was always highly supported (100%).

Trees were incongruent with regard to the central region. Whereas this clade split from the three others with well supported values in the slowly evolving cyt-b gene, it split off together with western individuals in a first step from eastern and island individuals in the faster evolving d-loop region.

3.5.1 Summary

As expected, the resolution of the highly variable d-loop region was higher than in the cyt-b gene, but tree topologies were not completely congruent for the position of central individuals.

The haplotype network did not provide a tracking of haplotype changes as none of the haplotypes was shared between the sample sites. Thus, the presented minimum spanning trees were randomly chosen from a set of all possible networks generated.

Overall, some general conclusions can be drawn. Central individuals were genetically different to the others, most distant to offshore island individuals and closest related to western individuals. Individuals from the offshore island were closer related to eastern individuals.

4. Discussion

The previously reported existence of two distinct morphotypes of *M. waterhousii* on Cuba (Gundlach, 1872; Rehn, 1904; Silva Taboada, 1979) is not supported by the findings of this study. There is no evidence for two separated populations on Cuba, neither in morphology, nor in echolocation, social interactions, or in genetic variability. Rather, all approaches resulted in a more complex pattern of variability. In the following, I shall first discuss the consequences of these patterns for potential colonisation scenarios and discuss drift versus positive selection as driving forces leading to these patterns. Finally, I develop a two-colonisation scenario with genetic drift as main force to account for the variability pattern of Cuban *M. waterhousii*.

Variability patterns and colonisation scenarios
Morphology
The characteristics of the subspecies *M. waterhousii minor* as described by Gundlach, 1864 do not fit the Cuban *M. waterhousii* population sampled in this study. The morphometrical descriptions instead matched measurements of western Cuban specimens only (forearm length of 47 mm, ear length of 25 mm and nose leaf length of 5.5 mm), but not those of the central, offshore island and eastern populations.

Although there was a general increase in forearm length and ear length of specimens from western to eastern Cuba, these size measures correlated only moderately with longitude. Other morphological parameters (nose leaf length, tragus width and snout width) did not change gradually. The most prominent parameter in this respect is snout width which influences bats´ ability to capture large prey and discriminated the four sample sites best. *Macrotus* is known to prey on large insects like katydids, dragonflies (Novick, 1963; Buden, 1975; Band, 2000), and also small vertebrates like lizards are commonly captured and eaten and even small bats (Dobson, 1878; Flower & Lydekker, 1891). Body size, aspect ratio of the wing and biting force (as a result of skull and mandible measurements in combination with muscular components) determine the size, hardness and weight of manageable prey (Fenton, 1990; Bogdanowicz et al., 1999; Sztencel et al., 2009). Snout width might influence the bat´s ability to handle caught prey items as bats must move large and struggling prey without letting go while eating. *M. waterhousii* specimens from the west have similar snout widths to those of the other sample sites, although they were smaller than the other specimens. Thus their prey handling ability is similar. The ecological consequences of narrow snouts of central individuals cannot be determined here. A difference of ~1 mm in snout width might not have sufficient disruptive power to limit access to different prey sources, especially as *M. waterhousii* has not specialized in specific insect species, although a difference of insect diversity in terms of species abundance in relation to biogeographic sectors of Cuba (Samek, 1973) has been reported and revealed comparatively low numbers in the central sector for single insect species or orders (Genaro & Tejuca, 1999; Peck, 2005; Genaro, 2006; Naranjo et al., 2010). A field study of the composition of insect species captured by *M. waterhousii* and experiments on biting force would provide tests of this hypothesis. As the pattern of variability is not continuously changing, the influence of an cline is unlikely. The only prominent ecological change across Cuba is a gradually increasing temperature eastwards (eastern area is on average 2.1°C warmer in annual average temperature).

A multiparametrical analysis revealed that specimens from the offshore island were similar to specimens from the east (Figure 21), although the sample sites were 485 km apart. Thus, the offshore island specimens may represent a group immigrated from the east, which used the chain of offshore islands as stepping stones. Western specimens were indeed the smallest ones in the morphological parameters with the exception of their snout width, which was as broad or broader than those of bats from the other sample sites. The area in the centre of Cuba did not represent intermediate forms in morphology compared to bats from the west or the east. Actually, specimens from this area were morphologically different from those of the other sample sites, differing mostly from the geographically closest offshore island specimens. The fact that western and eastern individuals were occasionally classified as central individuals according to their morphology, whereas central individuals were never classified as belonging to any of these groups, might be explained by a two-colonisation scenario rather than a single colonisation event in the central region. In the two-colonisation scenario, some individuals from the western founder population may have hybridized with individuals from the eastern founder population in the central region of Cuba. This small group of specimens then may have locally evolved the hybridized set of morphometric characteristics further into distinct character states.

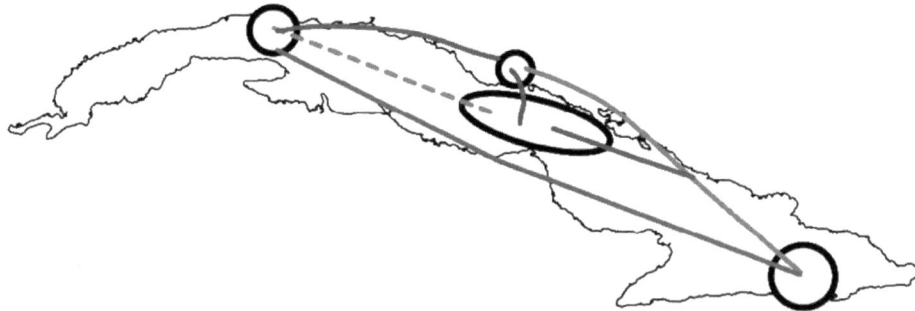

Figure 21: Variability pattern of morphology of M. waterhousii from four Cuban sample sites. Red lines indicate the existence of significant differences in morphology. Green and grey lines refer to non-significant sample site comparisons. They refer to an unambiguous (green), or ambiguous (grey) classification of individuals from the respective sample site comparison.

Echolocation

The effect of allometry on the acoustic characteristics of echolocation pulses was low and the acoustic variability structured the Cuban sample into three groups. In a standardized situation, echolocation pulses of specimens from the west and the central area were acoustically indistinguishable. The distinctiveness of echolocation pulses of bats from the east and from the offshore island was due to an emission, and the intensity of the fundamental. Eastern bats emitted it, but at low intensities, thus reducing the effective bandwidth of the pulse. As a result, peak frequencies differed as well. In pulses with a fundamental, the peak frequency was lower than in pulses with no or faint fundamental, where it shifts to higher frequencies.

From the two possible colonisation scenarios, namely a single colonisation event or the two-colonisation scenario, the pattern of acoustic characteristics across Cuban *M. waterhousii* (Figure 22) may again fit a two-colonisation scenario best. Whereas the western founder colony emitted echolocation pulses with a pronounced fundamental, the eastern colony, emitted pulses with only a faint fundamental and higher frequency. The fundamental might have been less frequently emitted while a group of eastern specimens moved along offshore islands, resulting in a complete loss of the fundamental in echolocation in the offshore island population. Maintenance of the fundamental in the pulses of bats in the central region, although they developed from western and eastern groups, might have been an evolutionary advantage as it results in a broadening of frequency range. A broad effective bandwidth results in an increased resolution of targets close to the background and influences the ability of bats to fly in highly cluttered habitats (Moss & Schnitzler, 1989; Fenton & Ratcliffe, 2004; Siemers & Schnitzler, 2004; Boonman & Ostwald, 2007; Schmieder *et al.*, 2010).

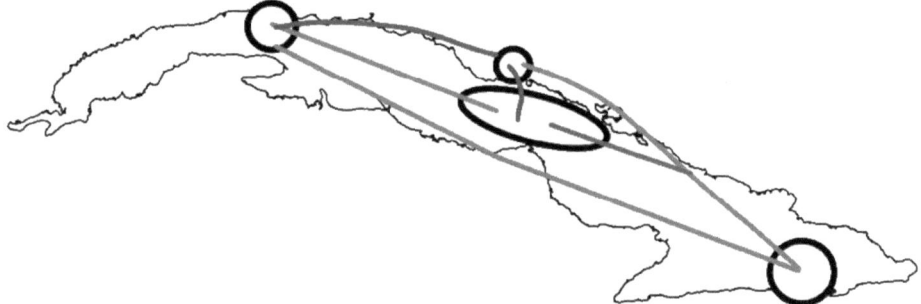

Figure 22: Variability pattern of echolocation pulse parameters of M. waterhousii from four Cuban sample sites. For further explanations, see Figure 21.

Genetic variability

The genetic variability of Cuban *M. waterhousii* among individuals from four samples sites and the haplotype analysis revealed, in general, low levels of diversity and no haplotypes were shared between the sample sites. As hypothesized by Fleming (Fleming, 1992), mobility is correlated with trophic position, which led to the assumption that genetic diversity should be low for insectivorous bats like *M. waterhousii*. In addition, those authors assume a low genetic diversity for Caribbean *M. waterhousii* because of its long period of residency in the Caribbean, its relatively small population size (in comparison to fruit eating bats) and its basal position in the phylogenetic tree of phyllostomids (Jones et al., 2002; Teeling et al., 2005). Indeed, Cuban *M. waterhousii* has a low genetic diversity and my results coincide with the findings of Fleming et al. (2009). Cuban *M. waterhousii* had a lower genetic diversity (0.0021, based on uncorrected p-distances) than mainland *M. waterhousii* (0.0031), Jamaican (0.0053) or *M. waterhousii* from Exuma (0.0096). However, compared to other Caribbean islands studied by Fleming et al., Cuban *M. waterhousii* had higher numbers of variable sites, number of haplotypes and haplotype diversity. The highest number of variable sites within their d-loop fragment was 15 for Jamaica compared to 42 in the Cuban sample. Haplotype diversity was also highest in Cuba with 0.97 and 16 haplotypes (N=15) compared to 0.53 and five haplotypes

(N=19 individuals) in Jamaica. They located 11 haplotypes across five Caribbean islands and thus 16 haplotypes on Cuba are more than expected. This could be a result of area size and/or colonisation history. The genetic structure of their sample revealed that islands are the major reason for structuring, as 95% of the variation occurred among islands (global FST= 0.953). In the Cuban sample, 76% of the variation occurred among the sample sites (global FST=0.755), with a low influence of geographic distance on genetic isolation (r^2=0.2 compared to r^2=0.6 in Fleming et al.´s sample). Fleming et al. assume at least four species in the *M. waterhousii* complex from their analyses, elucidating that *M. waterhousii* is much more sedentary then previously recognized and the rates of migration between islands are very low. This is strengthened by my analysis of the Cuban offshore island individuals. They are more than 200 mutational steps away from other Cuban haplotypes – a number which exceeds the maximum number of 88 mutational steps between Caribbean islands and the Mexican mainland (Grand Cayman and Sonora).

As a consequence of low genetic diversity within the Cuban sample, no clear scenario was obvious. Instead, the pattern was derived from genetic distances between the sample sites.

What can be concluded from both, the phylogenetic and haplotype analysis, is a high genetic similarity of eastern and offshore island individuals (Figure 23).

Furthermore, genetic differences between western and eastern individuals indicate dissimilarity. Buden (1975) interpreted the discontinuous variability of Cuban *M. waterhousii* as being a result of the genetic dilution of small Cuban bats with larger Hispaniolan bats which came across the windward passage to the east of Cuba. This assumption is not supported by my findings, as eastern individuals were genetically distant to central and western individuals. The fact that central individuals did not represent a mosaic of western and eastern individuals and were closer to western individuals than to eastern individuals, although incongruently placed within the phylogenetic trees, supports the assumption that they evolved distinct states of characters after a hybridization of western and eastern character sets with stronger influence of western characters. Therefore, the genetic dissimilarity between central and eastern individuals makes the scenario of a single colonisation event of the central region and/or the west or east unlikely and supports a two-colonisation scenario.

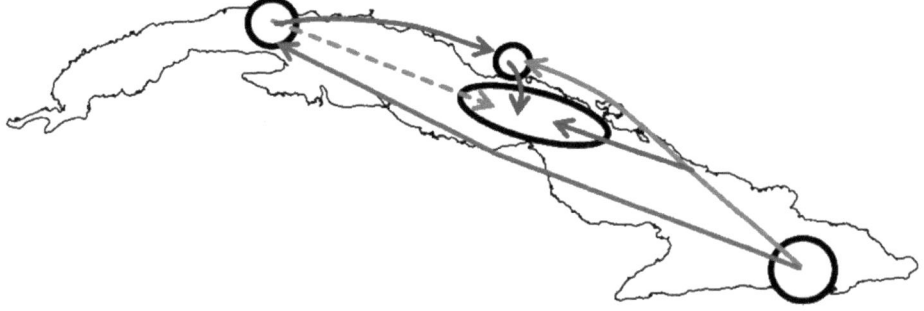

Figure 23: Variability pattern of combined phylogeny and haplotype results of cytochrome-b gene and d-loop marker of M. waterhousii from four Cuban sample sites. The dashed line indicates an incongruent result in phylogeny and haplotype analyses. For further explanations, see Figure 21.

Drift or positive selection?

Speciation is based on two main forces, genetic drift and positive selection. In the Isolation Concept, speciation is initiated by disruption of gene flow due to any gene-pool segregation, whereas speciation in the Selection Concept is driven by adaptations to different environments. Speciation then progresses in the Isolation Concept by genetic drift and/or founder effects, and/or natural selection, but processes in the Selection Concept by natural selection and/or sexual selection, which may act as a reinforcement to prezygotic isolation (Schilthuizen, 2000). I tested if *M. waterhousii* from different sites on Cuba differed in manoeuvrability, prey capture success under clutter conditions and in social interactions and if differences have the potential to act as barriers. Differences in manoeuvrability and echolocation behaviour may reflect adaptations into different microhabitats, thus create an ecological barrier, whereas restricted interactions between individuals from different sample sites refer to an ethological barrier.

Manoeuvrability and prey capture success in clutter

The experiment addressing manoeuvrability revealed that larger specimens from the central area and those from the east had to increase flight speed, but still collided in a two-obstacle course more often than western bats. Thus, differences in size had an effect on manoeuvrability. *M. waterhousii* is a generalist in terms of roost usage, and in addition, Cuba has a high number of caves. Thus, different manoeuvrability may not lead to diverging forces based on different roosting habitats. A different access to prey in different habitats due to size and manoeuvrability, however, may have the power to shift populations into different niches. Such a force is unlikely to exist in the Cuban *M. waterhousii* due to the previously mentioned general insect diet, in combination with the high ecological richness of Cuba.

Bats were exposed to a situation of foraging in clutter without passive acoustic cues. Western bats, which had a higher manoeuvrability and broader effective bandwidth, had the highest prey capture success rates in the sample site comparison. Although individuals from the central region were large in size and therefore had a lower manoeuvrability than western bats, prey capture success rate was high. They also emitted a pronounced fundamental, which indicates the advantage of a broad effective bandwidth and explains why central bats maintained the fundamental although originated from a hybridized group of specimens of western (pronounced fundamental) and eastern (faint fundamental) in the two-colonisation scenario. Consequently, eastern bats had lowest prey capture success rates.

Social interactions

The analysis of social call types studied here, revealed no clear differences between calls from bats of the different sample sites. This might be due to the small sample size. In addition, the recorded social calls were emitted in a context not suited to reveal sample site specific differences. Whether these calls serve to cry for help, to warn others or to appease, they inform the listeners of a high arousal situation. Such call types are not under positive selection pressure regarding individual or species specific signatures as, for instance, mating calls.

It is known from anecdotal reports in the literature that *Macrotus* males emit chirping sounds while they show a specific display mainly during mating season (Berry & Brown, 1995). Therefore studies focussing on communication between males and females and/or between males are promising for the analysis of ethological barriers.

Although I cannot reject the existence of a sample site-specific dialect in the call repertoire or in the behavioural display of Cuban *M. waterhousii* in general, I found no signs of isolating ethological barriers between bats from different sample sites. The affiliation among sample site members to allogroom with each other and to occupy specific areas in the flight room might be due to the fact that bats captured in an area were kept as a group inside a transportation cage for some days before being released into the flight room and not because of ethological differences between populations. This assumption is supported by the increase of allogrooming interactions of mixed-site dyads over time.

As no ecological or ethological barriers shaped by selection forces were obvious, variability in morphology, foraging behaviour, and echolocation are likely a result of genetic drift.

Which colonisation scenario for Cuba?

Islands of the Caribbean form a patchwork quilt in which each population responds independently to the particular environmental changes of its own location. This response of each population also depends on the particular pattern of new mutations and of new genetic combinations that have taken place within it. Therefore, each population will come to differ from the others. Thus, double invasions after becoming reproductively isolated in an allopatric phase can create a structure of variability within a continuous landmass.

Previous studies of *M. waterhousii* revealed that western Cuban *M. waterhousii* are more closely related to mainland *M. waterhousii* than to other Antillean *M. waterhousii* populations (Rehn, 1904; Koopman *et al.*, 1957; Anderson & Nelson, 1965; Buden, 1975). These small bats probably came directly from the Mexican mainland to Cuba. The larger eastern bats may have come as a second founder population from Hispaniola, where *M. waterhousii* specimens represent the largest of all forms. In this two-colonisation scenario (Figure 24), my findings indicate that western and eastern populations changed their characters states randomly by genetic drift effects. These two populations moved towards the centre of the island, both further diverging while moving, and met in the central area where they hybridized, locally developing distinct characteristics. This unique set of characteristics would be the result of receiving different drifted character states from the west and from the east, combining them and, in addition, continuously evolving these characteristics.

My study revealed that populations may indeed have faced a second contact on the island and Cuba was most likely colonised twice by *M. waterhousii* as proposed before.

Figure 24: Two-colonisation scenario including the results of this study. Arrows on the left and on the right indicate the two colonisation events as proposed in the literature. One brought M. waterhousii specimens directly from Yucantán to the western part of Cuba, whereas the other colonisation event brought specimens from southern Mexico and Guatemala via Jamaica and Hispaniola indirectly to the eastern part of Cuba. On Cuba, western specimens moved eastwards on the main island to the central region. Eastern specimens moved towards the centre along an offshore island chain. A few eastern specimens inhabited the central region on the main island and hybridized with western specimens.

The degree of variability does not support a taxonomic differentiation within Cuban *M. waterhousii*, but strengthened the assumption that *M. waterhousii* populations on different Caribbean islands may represent subspecies or species.

5. Appendix

5.1 Taxonomic classification of *Macrotus*

Table 25: Overview of taxonomic classifications within the genus Macrotus as proposed in the literature. Given is the taxonomic level, the species or subspecies name, its type locality and the respective authors. Table continued next page.

Taxonomic level	Name	Type locality	Author
GENUS	***Macrotus*** (=*"Otopterus"*)	Haiti	Gray, 1843 (Flower and Lydekker 1891)
SPECIES	*Macrotus waterhousii*	Haiti	Gray, 1843
SPECIES	*Macrotus californicus*	California	Baird, 1858
SPECIES	*Macrotus mexicanus*	Mexico, south-central	Saussure, 1860
SPECIES	*Macrotus minor*	Cuba, west	Gundlach, 1864
SPECIES	*Macrotus bocourtianus*	Guatemala	Dobson, 1876
SPECIES	*Macrotus bulleri*	Mexico, south-west	Allen, 1890
SPECIES	*Macrotus pygmeus*	Mexico, south-east	Rehn, 1904
	= *Micronycteris megalotis*		Goodwin, 1953
SPECIES	***Macrotus waterhousii,***		***Gray 1843***
SUBSPECIES	*M. w. jamaicensis*	Jamaica	
SUBSPECIES	*M. w. compressus*	Bahamas, north	Rehn, 1904
SUBSPECIES (SPECIES TO SUBSPECIES)	*M. w. minor*		Rehn, 1904

Table 25 continued.

SUBSPECIES	*M. w. heberfolium*	Bahamas, south	Shamel, 1931
SUBSPECIES (SP TO SSP)	*M. w. waterhousii*		Koopman, 1957
SUBSPECIES (SP TO SSP)	*M. w. bulleri*		Anderson and Nelson, 1965
SUBSPECIES (SP TO SSP)	*M. w. mexicanus*		Anderson and Nelson, 1965
SUBSPECIES (SP TO SSP)	*M. w. californicus*		Anderson and Nelson, 1965
SPECIES	***Macrotus mexicanus,***		***Saussure, 1860***
SUBSPECIES (SP TO SSP)	*M. m. bulleri*		Rehn 1904
SUBSPECIES (SP TO SSP)	*M. m. mexicanus*		Hall and Kelson, 1959

5.2 Distances between the sample sites

Table 26: Distances between the four sample areas in km.

Sites	Distance
west to central	317.74
west to island	298.63
west to east	726.81
central to island	125.63
central to east	456.80
island to east	485.20

Table 27: Cross-table of distances between the caves in km. Table continued next page.

Caves	Candela	Catalina 2	Catalina 3	Santo
Candela				
Catalina 2	10.31			
Catalina 3	11.04	0.79		
Santo	37.21	29.69	28.99	
Esperanza	286.30	277.94	277.53	277.45
Belleza	287.49	278.93	278.50	277.43
Salónes	321.61	312.70	312.23	309.10
Tres Ventiladores	336.04	327.16	326.68	323.60
Portales	444.81	436.17	435.72	433.54
Pelo de Oro	309.61	299.52	298.86	286.54
Mito	729.43	721.32	720.94	720.90
Bucanero	736.81	729.12	728.78	730.55
Terrarium	733.08	725.38	725.04	726.78
Costa	731.79	724.10	723.76	725.55
Juraguá	734.01	726.30	725.95	727.65

Table 27 continued.

Caves	Esperanza	Belleza	Salónes	Tres Ventiladores	Portales
Candela					
Catalina 2					
Catalina 3					
Santo					
Esperanza					
Belleza	9.50				
Salónes	45.37	38.85			
Tres Ventiladores	56.99	51.67	14.50		
Portales	159.28	157.37	124.65	110.20	
Pelo de Oro	124.30	114.89	100.96	107.13	191.87
Mito	443.66	443.47	413.54	399.21	289.36
Bucanero	453.31	454.08	426.83	412.83	304.93
Terrarium	449.54	450.29	423.03	409.02	301.13
Costa	448.32	449.10	421.91	407.92	300.10
Juraguá	450.39	451.11	423.77	409.75	301.77

Table 27 continued.

Caves	Pelo de Oro	Mito	Bucanero	Terrarium	Costa	Juraguá
Candela						
Catalina 2						
Catalina 3						
Santo						
Esperanza						
Belleza						
Salónes						
Tres Ventiladores						
Portales						
Pelo de Oro						
Mito	472.53					
Bucanero	493.33	46.02				
Terrarium	489.55	44.73	3.81			
Costa	488.65	45.60	5.03	1.63		
Juraguá	490.07	43.67	3.40	1.50	3.11	

5.3 Emission of the fundamental in echolocation pulses

Table 28: Overview of numbers of pulses with a prominent 2nd harmonic. Given are absolute numbers and respective percentages for all pulses.

Sample site	absolute number	%
Cuba (N=687)	566	82
West (N=184)	158	86
Central (N=169)	162	96
Offshore island (N=175)	101	58
East (N=159)	145	91

5.4 Photos of reproductive status of *M. waterhousii*

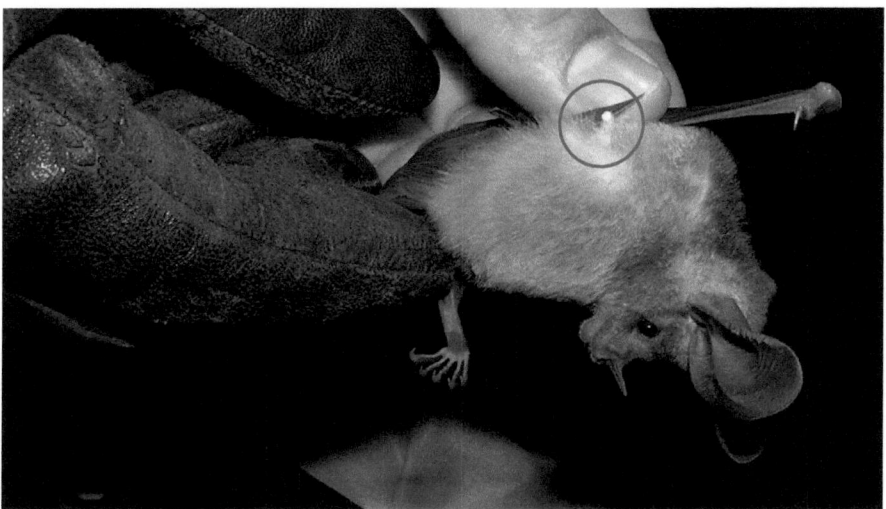

Figure 25: Lactating Macrotus female. Secretion of mammary gland see circle.

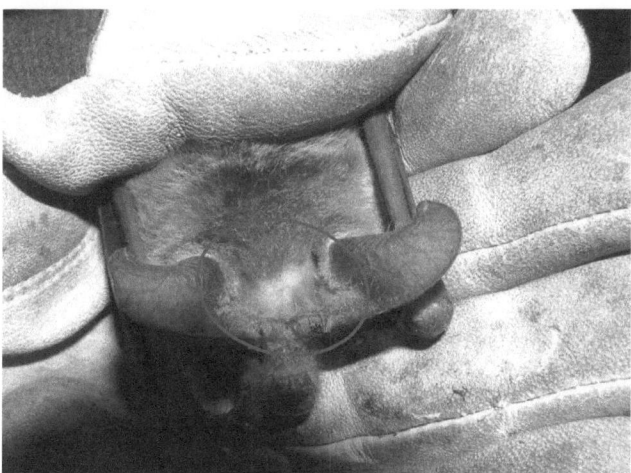

Figure 26: Bold-headed male Macrotus. Loss of fur was due to secretion of head glandulae. Secretion activity of glandulae was already reduced in this individual.

5.5 List of individuals kept in the lab

Table 29: List of bats kept in a facility at the University of Havana in the years 2008, and/or 2009/2010. Given are individual marks, sex, site at which the bat was captured, cave of capture and date of capture. Letters in bold refer to individuals which participated in experiments. Table continued next page.

Individual mark	Sex	Capture site	Colony	Date of capture
2008				
O	m	West	Catalina 2	01.06.2008
P	m	West	Catalina 2	01.06.2008
Q	m	West	Catalina 2	01.06.2008
R	f	West	Catalina 2	01.06.2008
S	f	West	Catalina 2	01.06.2008
A	m	East	La Terrarium	05.09.2008
D	f	East	La Terrarium	05.09.2008
C	m	East	La Terrarium	05.09.2008
F	f	East	La Terrarium	06.09.2008
H	m	East	La Terrarium	06.09.2008
I	f	East	La Terrarium	06.09.2008
K	m	East	Cueva La Costa	07.09.2008
L	m	East	Cueva La Costa	10.09.2008
N	f	East	Cueva La Costa	10.09.2008
2009/2010				
A	f	West	Cueva de la Candela	16.01.2010
B	f	West	Cueva de la Candela	16.01.2010
D	m	West	Cueva de la Candela	14.03.2010
E	m	West	Cueva de la Candela	14.03.2010

Table 29 continued.

H	m	West	Cueva del Santo	09.01.2010
J	m	Central	La Belleza	21.01.2010
K	m	Central	La Belleza	21.01.2010
N	f	Central	La Belleza	21.01.2010
O	f	Central	La Belleza	21.01.2010
P	f	Central	La Belleza	21.01.2010
Q = 2	f	East	El Mito	19.02.2010
R = 3	f	East	El Mito	19.02.2010
S	m	East	Hotel Bucanero	10.02.2010
T	m	East	El Mito	19.02.2010
U = 4	f	East	El Mitico	11.02.2010
V	f	East	El Mitico	11.02.2010
W = 5	m	East	El Mito	19.02.2010
X	f	East	El Mico	11.02.2010
Y = 6	m	East	Hotel Bucanero	02.02.2010
Z	f	East	El Mitico	11.02.2010
1	f	East	El Mito	19.02.2010
7	f	East	El Mito	19.02.2010

5.6 Photos of defined areas used in the experiment on spatial allocation

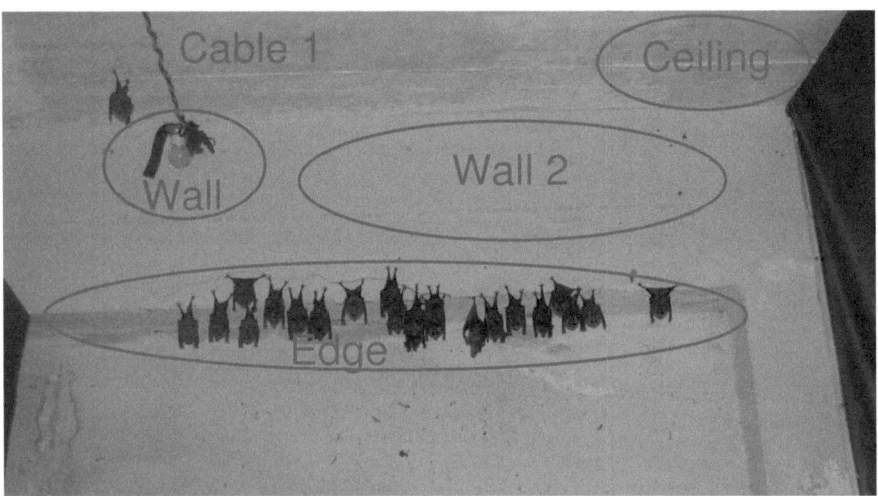

Figure 27: Photos of the defined areas inside the flight room. Respective areas are circled and named in red. Figure continued next page.

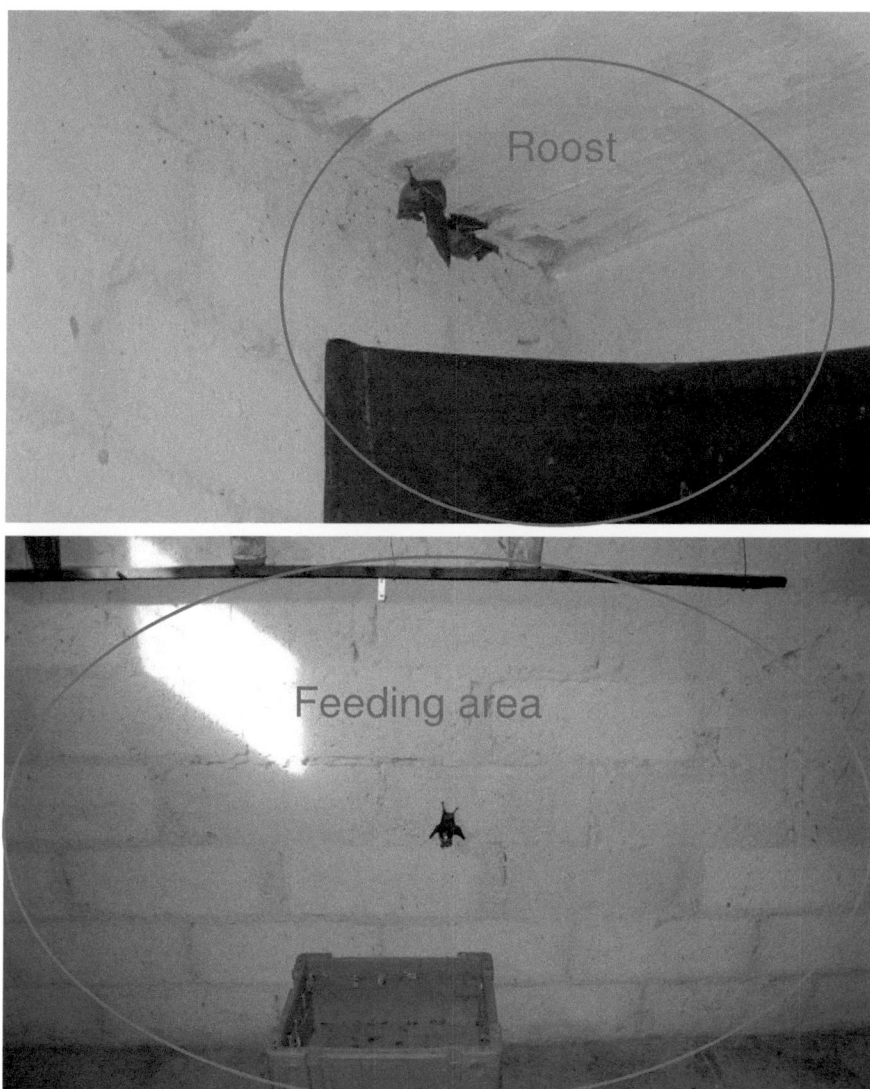

Figure 27 continued.

5.7 Analysed sessions in the experiment on passive listening

Table 30: Overview of analyzed sessions in the experiment on passive listening revealing an unbalanced data basis. Given are the frequencies showing how often each feeding dish (plastic, paper, metal) were presented at each position (left, middle, right) for each individual.

	Individual 2			Individual 5			Individual 6			Individual R		
	left	middle	right	left	middle	right	left	middle	right	left	middle	right
Plastic	3	3	3	1	5	3	3	2	4	4	1	3
Paper	3	2	4	3	3	3	1	5	3	4	1	3
Metal	3	4	2	5	1	3	5	2	2	0	6	2

5.8 Descriptive statistics for morphometric measurements

Table 31: Descriptive statistics of morphometric parameters measured in 77 adult individuals from four sample sites in Cuba. Given are medians, upper and lower quartile for each parameter. The first table presents data of all 77 individuals, followed by tables presenting data for western, central, offshore island and eastern individuals. Please note that the wing measurements are given only in the first table (based on 29 individuals. 7 western, 8 central, 3 offshore island and 11 eastern). Table continued next page.

Parameter	Median	75th Quartile	25th Quartile
Cuba (N=77)			
Forearm length (mm)	50.92	53.16	49.42
Nose leaf length (mm)	7.86	8.44	7.62
Nostril dist. (mm)	2.16	2.24	2.10
Tragus length (mm)	9.98	10.34	9.55
Tragus width (mm)	3.46	3.68	3.33
Snout width (mm)	2.56	3.28	2.40
Ear length (mm)	24.26	25.60	23.46
Wing span (cm)	31.20	32.78	30.75
Wing area (cm^2)	196.05	296.98	285.55
Aspect ratio	5.0	5.5	4.5
Wing load (Nm^{-2})	7.22	7.92	7.02
West (N=15)			
Forearm length (mm)	46.38	49.42	43.00
Nose leaf length (mm)	7.42	7.80	7.18
Nostril dist. (mm)	2.12	2.18	2.10
Tragus length (mm)	9.30	9.88	8.88
Tragus width (mm)	3.38	3.46	3.28
Snout width (mm)	3.00	3.42	2.46
Ear length (mm)	22.78	24.10	21.63

Table 31 continued.

Central (N=33)

Forearm length (mm)	50.30	51.18	49.60
Nose leaf length (mm)	8.44	8.70	7.90
Nostril dist. (mm)	2.14	2.22	2.06
Tragus length (mm)	10.17	10.80	9.40
Tragus width (mm)	3.42	3.84	3.24
Snout width (mm)	2.42	2.56	2.26
Ear length (mm)	23.72	24.36	23.46

Offshore island (N=13)

Forearm length (mm)	54.69	56.08	53.29
Nose leaf length (mm)	7.62	7.64	7.60
Nostril dist. (mm)	2.20	2.24	2.17
Tragus length (mm)	9.90	10.18	9.84
Tragus width (mm)	3.56	3.68	3.42
Snout width (mm)	3.28	3.29	3.28
Ear length (mm)	26.26	26.76	25.77

East (N=16)

Forearm length (mm)	53.31	59.57	51.24
Nose leaf length (mm)	8.08	8.21	7.68
Nostril dist. (mm)	2.22	2.34	2.10
Tragus length (mm)	10.10	10.37	9.77
Tragus width (mm)	3.63	3.69	3.49
Snout width (mm)	2.78	3.30	2.55
Ear length (mm)	26.51	27.84	25.00

5.9 Correlation between morphometrical parameters

Table 32: Spearman correlation coefficients between the morphological parameters.

	forearm length	nose leaf length	nostril distance	tragus length	tragus width	snout width	ear length
nose leaf length	0.1						
nostril distance	0.3	-0.1					
tragus length	0.1	0.2	0.2				
tragus width	0.4	-0.0	0.2	0.2			
snout width	0.4	-0.3	0.3	-0.1	0.2		
ear length	0.6	-0.1	0.3	0.1	0.3	0.3	

5.10 Multiparametrical comparison of the four sample sites based on morphology

Table 33: Results of multiple group comparison. The four sampling sites were compared by their morphological parameters with exclusion of outliers and extremes (West: N = 15, Central: N = 33, Offshore island: N = 16, East: N = 13). Given is the Kruskal Wallis H for each parameter. Critical Chi^2 (d.f.=3, N=77) =7.81. Level of significance 5%, significant results are marked by asterisks.

Parameter	Kruskal Wallis H
Forearm length	44.99 *
Nose leaf length	28.22 *
Nostril distance	7.01
Tragus length	11.88 *
Tragus width	7.67
Snout width	34.39 *
Ear length	38.24 *

Table 34: Significant results of multiple pairwise comparisons of morphological parameters of individuals from four sample sites. Level of significance was Bonferroni corrected (= ≤0.008).

West versus central	Rank Sum	Rank Sum	U	Z	p-level
forearm length	333.5	1151.5	123.5	-3.878	0.000
nose leaf length	314.0	1171.0	104.0	-4.228	0.000
tragus length	364.0	1121.0	154.0	-3.332	0.001
snout width	742.0	743.0	148.0	3.445	0.000
West versus offshore island					
forearm length	214.0	416.0	4.0	-4.867	0.000
tragus length	252.5	377.5	42.5	-3.586	0.000
tragus width	274.5	355.5	64.5	-2.854	0.003
West versus east					
forearm length	226.0	554.0	16.0	-4.889	0.000
nose leaf length	306.0	474.0	96.0	-2.641	0.008
tragus length	262.5	517.5	52.5	-3.865	0.000
tragus width	277.5	502.5	67.5	-3.445	0.000
Central versus offshore island					
forearm length	606.0	619.0	24.0	5.011	0.000
nose leaf length	204.0	1021.0	84.0	-3.710	0.000
snout width	630.0	595.0	0.0	5.541	0.000
Central versus east					
forearm length	676.0	755.0	81.0	-4.489	0.000
nose leaf length	1102.5	328.5	138.5	3.423	0.000
snout width	672.5	758.5	77.5	-4.562	0.000
Offshore island versus east					
nose leaf length	165.0	430.0	45.0	-3.382	0.000

Table 35: Significant results of MANOVA post hoc tests of multiple pairwise comparisons per morphometrical parameter (first column). Name of the test (second column) is followed by sample site comparison pair. Mean difference between respective sample site pair are followed by test results. Table continued on next page.

Parameter	Test	Site	Versus site	Mean Diff.	Std. Error	p-value	Conf. interv. 95 % lower bound.	Conf. interv. 95 % upper bound.
Forearm length	Gabriel	central	east	-1.167	0.207	0.000	-1.717	-0.618
			island	-0.985	0.222	0.000	-1.570	-0.401
			west	0.901	0.211	0.000	0.341	1.461
		east	west	2.068	0.244	0.000	1.410	2.727
		island	west	1.887	0.257	0.000	1.193	2.580
	Games-Howell	central	east	-1.167	0.285	0.004	-1.982	-0.352
			island	-0.985	0.138	0.000	-1.379	-0.592
			west	0.901	0.216	0.004	0.283	1.520
		east	west	2.068	0.349	0.000	1.113	3.024
		island	west	1.887	0.245	0.000	1.209	2.565
Nose leaf length	Gabriel	central	island	1.050	0.274	0.001	0.329	1.771
			west	1.433	0.260	0.000	0.743	2.124
	Games-Howell	central	east	0.641	0.186	0.006	0.146	1.135
			island	1.050	0.249	0.002	0.364	1.736
			west	1.433	0.335	0.002	0.498	2.369
Tragus length	Gabriel	central	west	1.055	0.289	0.002	0.289	1.822
		east	west	1.071	0.334	0.012	0.169	1.972
	Games-Howell	central	west	1.055	0.338	0.021	0.131	1.980
		east	west	1.071	0.325	0.016	0.172	1.969

Table 35 continued

		island	west	0.903	0.294	0.032	0.067	1.739
	Games-Howell	east	west	0.901	0.216	0.004	0.283	1.520
Snout width	Gabriel	central	east	-0.901	0.216	0.004	-1.520	-0.283
			island	-2.068	0.349	0.000	-3.024	-1.113
			west	-1.887	0.245	0.000	-2.565	-1.209
		east	island	1.050	0.274	0.001	0.329	1.771
	Games-Howell	central	east	-1.433	0.260	0.000	-2.124	-0.743
			island	-0.793	0.301	0.059	-1.604	0.019
			west	-0.383	0.317	0.784	-1.238	0.472
		east	island	1.050	0.249	0.002	0.364	1.736
Ear length	Gabriel	central	east	-1.433	0.335	0.002	-2.369	-0.498
			island	-0.793	0.324	0.103	-1.707	0.122
		east	west	-0.073	0.302	1.000	-0.873	0.726
		island	west	0.532	0.348	0.561	-0.408	1.472
	Games-Howell	central	east	0.073	0.302	1.000	-0.726	0.873
			island	-0.532	0.348	0.561	-1.472	0.408
		east	west	-0.073	0.220	0.987	-0.659	0.512
		island	west	0.532	0.283	0.264	-0.251	1.314

5.11 Classification of individuals based on morphology

Table 36: Classification results of the discriminant analysis on 77 individuals based on median morphological characteristics. 70.1% of cross-validated individuals were correctly classified to their respective sample site.

Sample site	West	Central	Offshore island	East	Total
West	60%	33%	7%	0%	100%
	(9)	(5)	(1)	(0)	(15)
Central	3%	97%	0%	0%	100%
	(1)	(32)	(0)	(0)	(33)
Offshore island	8%	8%	84%	0%	100%
	(1)	(1)	(11)	(0)	(13)
East	0%	55%	32%	13%	100%
	(0)	(9)	(5)	(2)	(16)

5.12 Flight speed of bats in the experiment on manoeuvrability

Table 37: Flight speed of 14 individuals in three experimental conditions. The first column indicates the individual and respective experimental condition marked by numbers 0 = no obstacle, 1= one obstacle and 2 = two obstacles. Given are number of analysed trials, median flight speed and lower and upper quartiles in m/s. The last column shows significant results of pair wise comparisons, Wilcoxon tests, calculated subsequent to significant results of Friedman ANOVA (d.f. = 2, p = < 0.05). Given is the Z value and the respective p-value (Bonferroni corrected = 0.025). "n.s." refers to non-significant results. Table continued next pages.

Ind	n	Median	25th Quartile	75th Quartile	Wilcoxon test
A 0	4	0.89	0.87	0.93	n.s.
A 1	5	0.93	0.80	1.00	n.s
A 2	7	1.02	0.95	1.09	n.s
B 0	5	0.67	0.60	0.73	n.s
B 1	5	0.93	0.93	0.93	n.s
B 2	6	1.20	1.16	1.38	n.s
E 0	5	1.07	1.07	1.07	n.s
E 1	5	0.87	0.87	0.87	n.s
E 2	6	1.02	1.02	1.02	n.s
H 0	5	0.67	0.67	0.80	n.s
H 1	5	0.87	0.87	0.87	n.s
H 2	5	1.09	1.02	1.24	n.s
J 0	5	0.88	0.87	0.90	n.s
J 1	6	0.91	0.87	0.93	Z = 2.201, p = 0.024
J 2	6	1.15	1.09	1.18	

Table 37 continued.

K 0	5	0.88	0.87	0.92	n.s
K 1	5	1.02	0.98	1.03	n.s
K 2	5	1.16	1.15	1.24	n.s
N 0	5	0.97	0.92	0.98	n.s
N 1	6	0.94	0.92	0.95	$Z = 2.201, p = 0.024$
N 2	7	1.20	1.09	1.31	
O 0	7	0.95	0.92	0.95	n.s
O 1	7	0.92	0.87	0.97	$Z = 2.366, p = 0.018$
O 2	7	1.27	1.20	1.31	
P 0	5	0.92	0.92	0.93	n.s
P 1	6	0.91	0.87	0.92	$Z = 2.201, p = 0.024$
P 2	6	1.25	1.15	1.31	
S 0	5	0.93	0.80	0.93	n.s
S 1	6	1.03	1.00	1.07	n.s
S 2	7	1.16	1.09	1.16	n.s
T 0	5	0.87	0.87	0.93	n.s
T 1	5	1.07	1.00	1.07	n.s
T 2	6	1.31	1.16	1.38	n.s
V 0	5	1.00	0.87	1.07	n.s
V 1	4	0.87	0.80	0.93	n.s
V 2	6	1.09	1.09	1.16	n.s

Table 37 continued.

X 0	5	0.93	0.93	1.07	n.s
X 1	7	0.93	0.87	0.93	$Z = 2.201, p = 0.024$
X 2	6	1.09	1.02	1.09	
Z 0	5	1.00	0.93	1.07	n.s
Z 1	5	0.80	0.73	0.80	$Z = 2.201, p = 0.024$
Z 2	6	0.98	0.95	1.02	

5.13 Descriptive statistics of echolocation pulse measurements

Table 38: Descriptive statistics for echolocation pulses of 77 adult individuals from four sampling sites in Cuba. Given are grand medians, upper and lower quartiles for each acoustic parameter. Upper part of the table presents data of all 77 individuals, followed by tables presenting the data for western, central, offshore island and eastern individuals on following pages.

Parameter	Grand median	75th Quartile	25th Quartile
Cuba (N=77)			
DurP (ms)	0.91	1.02	0.75
PeakfP (kHz)	73.33	78.75	68.93
TermfP (kHz)	32.91	46.71	30.24
StartfP (kHz)	109.68	112.25	106.54
eff.BandwP (kHz)	76.27	80.65	62.99
BandwP (kHz)	76.70	81.17	71.84
ModRate (kHz/ms)	81.37	97.79	71.76
2ndHarmPeakf (kHz)	68.55	71.58	64.95
2ndHarmTermf (kHz)	53.95	56.87	49.75
2ndHarmStartf (kHz)	84.58	88.22	79.57
2ndHarmBandw (kHz)	30.44	33.73	28.15
2ndHarmModRate (kHz/ms)	32.80	40.61	29.94
West (N=15)			
DurP (ms)	0.89	1.03	0.73
PeakfP (kHz)	71.94	75.27	68.87
TermfP (kHz)	30.78	32.52	27.75
StartfP (kHz)	110.04	112.66	107.43
eff.BandwP (kHz)	78.00	81.52	76.18
BandwP (kHz)	78.00	81.52	76.18
ModRate (kHz/ms)	88.12	110.33	75.82

Table 38 continued.

2^{nd}HarmPeakf (kHz)	70.00	73.93	67.67
2^{nd}HarmTermf (kHz)	56.21	58.20	54.42
2^{nd}HarmStartf (kHz)	86.01	89.32	81.39
2^{nd}HarmBandw (kHz)	30.97	33.95	24.21
2^{nd}HarmModRate (kHz/ms)	31.81	34.05	28.31
Central (N=33)			
DurP (ms)	0.95	0.85	1.12
PeakfP (kHz)	73.48	70.36	79.88
TermfP (kHz)	32.59	30.28	36.49
StartfP (kHz)	110.10	106.15	112.57
eff.BandwP (kHz)	76.89	81.05	66.71
BandwP (kHz)	76.89	66.71	81.05
ModRate (kHz/ms)	79.04	58.00	89.84
2^{nd}HarmPeakf (kHz)	71.27	68.55	73.96
2^{nd}HarmTermf (kHz)	55.48	53.29	57.66
2^{nd}HarmStartf (kHz)	87.56	86.12	91.64
2^{nd}HarmBandw (kHz)	32.97	29.27	35.41
2^{nd}HarmModRate (kHz/ms)	32.86	27.66	40.07
Offshore island (N=13)			
DurP (ms)	0.71	0.65	0.82
PeakfP (kHz)	79.23	77.00	81.77
TermfP (kHz)	47.03	46.48	49.13
StartfP (kHz)	110.42	108.90	112.53
eff.BandwP (kHz)	61.85	57.25	65.95
BandwP (kHz)	62.88	57.95	66.50

Table 38 continued.

ModRate (kHz/ms)	82.73	71.61	98.27
2ndHarmPeakf (kHz)	63.53	62.38	65.55
2ndHarmTermf (kHz)	47.26	45.29	48.51
2ndHarmStartf (kHz)	75.45	72.60	80.27
2ndHarmBandw (kHz)	28.64	26.31	30.44
2ndHarmModRate (kHz/ms)	39.55	30.35	46.35
East (N=16)			
DurP (ms)	0.87	0.73	0.98
PeakfP (kHz)	68.61	63.94	74.58
TermfP (kHz)	44.09	28.53	50.06
StartfP (kHz)	106.23	103.78	108.40
eff.BandwP (kHz)	63.26	54.80	79.97
BandwP (kHz)	74.88	72.61	81.35
ModRate (kHz/ms)	77.00	68.51	97.95
2ndHarmPeakf (kHz)	64.11	62.23	66.21
2ndHarmTermf (kHz)	50.12	47.66	51.30
2ndHarmStartf (kHz)	79.26	76.20	81.64
2ndHarmBandw (kHz)	29.20	27.70	31.30
2ndHarmModRate (kHz/ms)	34.52	28.36	40.61

5.14 Correlations between the acoustical parameters

Table 39: Spearman correlation coefficients between the morphological measures and the acoustic measures.

Correlations	DurP	PeakP	TermF	eff. BandwP	Peak2ndHarm	2ndHarm TermF
Forearm length	-0.23	0.12	0.52	-0.47	-0.54	**-0.66**
Nose leaf length	0.26	-0.05	-0.12	0.09	0.27	0.17
Nostril distance	-0.19	0.10	0.17	-0.08	-0.13	-0.15
Tragus length	-0.04	-0.01	0.22	-0.32	-0.05	0.02
Tragus width	-0.11	0.05	0.12	-0.08	-0.11	-0.14
Snout width	-0.43	0.22	0.31	-0.37	-0.38	-0.35
Ear length	-0.22	0.06	0.46	-0.42	-0.51	**-0.66**

5.15 Multiparametrical comparisons of the four sample sites based on echolocation

Table 40: Results of multiple group comparison. The four sampling sites were compared by their median call parameters (west: N=15, central: N=33, island: N= 13, east: N= 16). Given is the Kruskal Wallis H for each parameter, correlating parameters excluded. Critical Chi² (d.f.= 3, N = 77) = 7.81. Level of significance 5%, significant results are marked by asterisks.

Parameter	Kruskal Wallis H
DurP	17.09
PeakfP	17.480 *
TermfP	19.35 *
eff.BandwP	15.27 *
2ndHarmPeakf	36.72 *
2ndHarmTermf	47.76 *

Table 41: Significant results of multiple pairwise comparisons of acoustic parameters of individuals from four sample sites. Level of significance was Bonferroni corrected (p=0.008). Table continued next page.

West versus central	Rank Sum	Rank Sum	U	Z	p-level
West versus offshore island					
PeakfP	132.0	274.0	12.0	-3.939	0.000
TermfP	120.0	286.0	0.0	-4.491	0.000
eff. BandwP	310.0	96.0	5.0	4.261	0.000
BandwP	314.0	92.0	1.0	4.445	0.000
2^{nd}HarmPeakf	300.0	106.0	15.0	3.800	0.000
2^{nd}HarTermf	315.0	91.0	0.0	4.491	0.000
2^{n}HarmStartf	309.0	97.0	6.0	4.215	0.000
West versus east					
2^{nd}HarmPeakf	316.0	180.0	44.0	3.004	0.003
2^{nd}HarmTermf	343.0	153.0	17.0	4.071	0.000
2^{nd}HarmStartf	337.0	159.0	23.0	3.834	0.000
Central versus offshore island					
DurP	939.0	142.0	51.0	3.989	0.000
TermfP	613.0	468.0	52.0	-3.964	0.000
eff. BandwP	907.0	174.0	83.0	3.208	0.001
Bandw	977.0	104.0	13.0	4.916	0.000
2^{nd}HarmPeakf	978.0	103.0	12.0	4.940	0.000
2^{nd}HarmTermf	987.0	94.0	3.0	5.160	0.000
2^{nd}HarmStartf	978.0	103.0	12.0	4.940	0.000
2^{nd}HarmBandw	885.0	196.0	105.0	2.671	0.008
DurP	939.0	142.0	51.0	3.989	0.000

Table 41 continued.

Central versus east

2^{nd}HarmPeakf	1027.0	198.0	62.0	4.307	0.000
2^{nd}HarmTermf	1047.0	178.0	42.0	4.733	0.000
2^{nd}HarmStartf	1060.0	165.0	29.0	5.010	0.000
2^{nd}HarmBandw	954.0	271.0	135.0	2.750	0.006

Offshore island versus east

PeakfP	271.0	164.0	28.0	3.333	0.001
Bandw	92.0	343.0	1.0	-4.517	0.000
2^{nd}HarmTermf	117.0	318.0	26.0	-3.421	0.001

Table 42: MANOVA post hoc test results of multiple pairwise comparisons of acoustic parameters of individuals from four sample sites. Given is the calculated mean difference of each parameter. In parenthesis respective p values of Gabriel´s test and Games-Howell test are given. Level of significance was Bonferroni corrected (p=0.008). "n.s." refers to test results which were not significant.

Parameter	West vs. central	West vs. island	West vs. east	Central vs. island	Central vs. east	Island vs. east
DurP	n.s.	n.s.	n.s.	n.s.	n.s.	n.s.
PeakfP	n.s.	-1.138 (0.005; 0.000)	n.s.	n.s.	n.s.	1.165 (0.003; 0.000)
TermfP	n.s.	-2.279 (0.000; 0.000)	n.s.	-2.031 (0.000; 0.000)	0.413 (0.001; 0.007)	2.444 (0.000; 0.000)
eff.Bandw	n.s.	1.179 (0.003; 0.000)	n.s.	0.894 (n.s; 0.001)	n.s.	n.s.
2nd HarmPeakf	n.s.	1.324 (0.000; 0.000)	0.977 (0.003; 0.008)	1.629 (0.000; 0.000)	1.282 (0.000; 0.000)	n.s.
2nd HarmTermf	n.s.	2.169 (0.000; 0.000)	1.408 (0.000; 0.000)	1.846 (0.000; 0.000)	1.085 (0.000; 0.000)	-0.761 (n.s.; 0.001)

5.16 Additional results of the experiment on prey capture success in clutter

Table 43: Results of Chi² statistics for approaches to dishes without subsequent landing per individual. Given are observed and expected frequencies and respective Chi² (d.f. = 2 for overall chi² and d.f. = 1 for single cells). Left side of the table gives frequencies of approaches without subsequent landings for positions, right part of the table for the materials. Significant deviations from a equal distribution of approaches without subsequent landings are marked by asterisks. A statistical trend is indicated by an asterisk in parentheses.

	Position	Obs.	Exp.	Chi²		Material	Obs.	Exp.	Chi²
Individual 2	left	8	12	1.330	Individual 2	Plastic	11	12	0.083
	middle	12	12	0.000		Paper	8	12	1.330
	right	16	12	1.330		Metal	17	12	2.083
	Σ	36		2.670		Σ	36		3.496
Individual 5	left	23	18.3	1.207	Individual 5	Plastic	17	18.3	0.923
	middle	14	18.3	1.010		Paper	15	18.3	0.595
	right	18	18.3	0.005		Metal	23	18.3	1.207
	Σ	55		2.222		Σ	55		2.725
Individual 6	left	24	17	2.882	Individual 6	Plastic	11	17	2.118
	middle	16	17	0.059		Paper	9	17	*3.765(*)*
	right	11	17	2.118		Metal	31	17	*11.529**
	Σ	51		5.059		Σ	51		*17.412**
Individual R	left	2	3.67	0.760	Individual R	Plastic	8	3.67	5.109
	middle	2	3.67	0.760		Paper	1	3.67	1.943
	right	7	3.67	3.022		Metal	2	3.67	0.760
	Σ	11		4.541		Σ	11		7.812

Table 44: Results of Chi² statistics for approaches to dishes with subsequent landing per individual. Given are observed and expected frequencies and respective Chi² (d.f. = 2 for overall chi² and d.f. = 1 for single cells). Left side of the table gives frequencies of approaches with subsequent landings for positions, right part of the table for the materials. Significant deviations from a equal distribution of approaches with subsequent landings are marked by asterisks.

	Position	Obs.	Exp.	Chi²		Material	Obs.	Exp.	Chi²
Individual 2	left	22	29.3	1.819	Individual 2	Plastic	41	29.3	**4.672***
	middle	24	29.3	0.959		Paper	38	29.3	2.583
	right	42	29.3	**5.505***		Metal	9	29.3	**14.065***
	Σ	88		**8.282***		Σ	88		**21.320***
Individual 5	left	16	11.4	1.856	Individual 5	Plastic	14	11.4	0.591
	middle	9	11.4	0.505		Paper	12	11.4	0.032
	right	9	11.4	0.505		Metal	8	11.4	1.014
	Σ	34		2.867		Σ	34		1.639
Individual 6	left	19	18.3	0.027	Individual 6	Plastic	21	18.3	0.398
	middle	16	18.3	0.289		Paper	16	18.3	0.289
	right	20	18.3	0.158		Metal	18	18.3	0.005
	Σ	55		0.474		Σ	55		0.692
Individual R	left	10	7.33	0.973	Individual R	Plastic	14	7.33	**6.069***
	middle	2	7.33	3.876		Paper	8	7.33	0.061
	right	10	7.33	0.973		Metal	0	7.33	**7.330***
	Σ	22		**5.822(*)**		Σ	22		**13.461***

Table 45: Frequencies of first approaches to previous reward field and respective p-value of a binomial test of each individual. Significant results are marked by asterisks.

Individual	Number of approaches to previous reward field	p-value
West		
A	14 of 18	p = 0.0154 *
B	12 of 18	p = 0.1189
E	16 of 18	p = 0.0007 *
H	6 of 6	p = 0.0001 *
Central		
J	16 of 18	p = 0.0007 *
K	13 of 18	p = 0.0481
N	11 of 18	p = 0.2403
O	14 of 18	p = 0.0154 *
P	15 of 18	p = 0.0038 *
East		
S	11 of 18	p = 0.2403
T	15 of 18	p = 0.0038 *
V	15 of 18	p = 0.0038 *
X	17 of 18	p = 0.0001 *
Z	16 of 18	p = 0.0007 *

Table 46: Number of hovering events prior landings and p-value of respective binomial test of each individual assuming that hovering occurred with equal probability. Significant results are marked by asterisks.

Individual	Number of hovering events	p value
West		
A	98 of 270	p = 0.0001 *
B	210 of 270	p = 0.0001 *
E	187 of 270	p = 0.0001 *
H	88 of 90	p = 0.0001 *
Central		
J	178 of 270	p = 0.0001 *
K	252 of 270	p = 0.0001 *
N	201 of 270	p = 0.0001 *
O	75 of 270	p = 0.0001 *
P	147 of 270	p = 0.0807
East		
S	98 of 270	p = 0.0001 *
T	236 of 270	p = 0.0001 *
V	213 of 270	p = 0.0001 *
X	199 of 270	p = 0.0001 *
Z	263 of 270	p = 0.0001 *
total	2578 of 3600	p = 0.0001 *

Table 47: Number of approaches to reward field and non-reward field in three clutter conditions of each individual. Last two columns give statistical results of Wilcoxon matched pairs test (level of significance was Bonferroni corrected) after a Friedman ANOVA revealed a significant difference between the three experimental conditions per individual (Chi^2 = 8.40, d.f. = 2, p = < 0.015, Coefficient of concordance = 0.840, average rank = 0.80). Test values were calculated as (Obs x Exp^2)/exp. Table continued next pages.

West

A	reward field	no reward field	Test value reward	Test value no reward
No clutter	281	10	n.s.	n.s.
Moderate clutter	290	19	n.s.	n.s.
High clutter	187	100	n.s.	n.s.
B	reward field	no reward field	Test value reward	Test value no reward
No clutter	279	3	n.s.	n.s.
Moderate clutter	281	7	n.s.	n.s.
High clutter	209	69	n.s.	n.s.
E	reward field	no reward field	Test value reward	Test value no reward
No clutter	290	5	n.s.	n.s.
Moderate clutter	283	7	n.s.	n.s.
High clutter	242	46	n.s.	n.s.

Table 47 continued.

H	reward field	no reward field	Test value reward	Test value no reward
No clutter	96	4	n.s.	n.s.
Moderate clutter	88	11	n.s.	n.s.
High clutter	62	49	n.s.	n.s.

Central

J	reward field	no reward field	Test value reward	Test value no reward
No clutter	285	9	n.s.	n.s.
Moderate clutter	288	15	n.s.	n.s.
High clutter	254	85	n.s.	n.s.

K	reward field	no reward field	Test value reward	Test value no reward
No clutter	293	7	n.s.	n.s.
Moderate clutter	301	19	n.s.	n.s.
High clutter	250	79	n.s.	n.s.

N	reward field	no reward field	Test value reward	Test value no reward
No clutter	289	4	n.s.	n.s.
Moderate clutter	273	21	n.s.	n.s.
High clutter	280	43	n.s.	n.s.

Table 47 continued.

O	reward field	no reward field	Test value reward	Test value no reward
No clutter	287	4	n.s.	n.s.
Moderate clutter	288	11	n.s.	n.s.
High clutter	297	35	n.s.	n.s.

P	reward field	no reward field	Test value reward	Test value no reward
No clutter	301	15	n.s.	n.s.
Moderate clutter	312	23	n.s.	n.s.
High clutter	288	69	n.s.	n.s.

East

S	reward field	no reward field	Test value reward	Test value no reward
No clutter	283	3	8.93	231.04
Moderate clutter	291	5	12.30	227.11
High clutter	131	181	47.41	13.23

T	reward field	no reward field	Test value reward	Test value no reward
No clutter	289	6	11.41	225.15
Moderate clutter	273	31	5.47	179.05
High clutter	89	219	92.42	1.37

Table 47 continued.

V	reward field	no reward field	Test value reward	Test value no reward
No clutter	291	21	12.30	196.86
Moderate clutter	174	57	16.75	136.71
High clutter	107	213	71.31	2.43

X	reward field	no reward field	Test value reward	Test value no reward
No clutter	203	99	4.88	80.35
Moderate clutter	149	56	32.68	138.23
High clutter	138	200	41.35	5.78

Z	reward field	no reward field	Test value reward	Test value no reward
No clutter	296	10	14.69	217.42
Moderate clutter	266	33	3.55	175.59
High clutter	89	239	92.42	0.02

Table 48: Number of landings in the reward and non-reward field in three clutter conditions of each individual. Observed number of landings was compared with expected number of landings. Expected number for a total of 270 landings in the reward and the non-reward field per clutter condition= 237 and for a total of 90 landings ("H") = 74. Results of Chi² tests are given in the last two columns which were compared with critical value of overall Chi² (d.f. = 2) = 5.991 and for single cells = 3.841. Significant results are marked by asterisks and arrows indicate if observed number of landings was higher (↑), or lower (↓), than expected. Asterisks at individual code refers to a significant result of overall Chi² test. Fields of test values remain empty if number of observed frequencies was zero. Table continued next pages.

	reward	no reward	Chi² rew (obs - exp)²/exp	Chi² no-rew (obs - exp)²/exp
West				
A *				
No clutter	270	0	4.59 * ↑	
Moderate clutter	270	0	4.59 * ↑	
High clutter	171	99	18.38 * ↓	80.35 * ↓
B				
No clutter	270	0	4.59 * ↑	
Moderate clutter	270	0	4.59 * ↑	
High clutter	207	63	3.80	127.75 * ↓
E				
No clutter	270	0	4.59 * ↑	
Moderate clutter	266	4	3.55	229.07 * ↓
High clutter	236	34	0.00	173.88 * ↓

Table 48 continued.

H

No clutter	90	0	3.46	
Moderate clutter	82	8	0.86	58.86 * ↓
High clutter	51	39	7.15 * ↓	16.55 * ↓

Central

J

No clutter	270	0	4.59 * ↑	
Moderate clutter	267	3	3.80 * ↑	231.04 * ↓
High clutter	224	46	0.71	153.93 * ↓

K

No clutter	270	0	4.59 * ↑	
Moderate clutter	270	0	4.59 * ↑	
High clutter	211	59	2.85	133.69 * ↓

N

No clutter	270	0	4.59 * ↑	
Moderate clutter	256	15	1.52	207.95 * ↓
High clutter	249	21	0.61	196.86 * ↓

Table 48 continued.

O

No clutter	270	0	4.59 * ↑	
Moderate clutter	270	0	4.59 * ↑	
High clutter	253	17	1.08	204.22 * ↓

P

No clutter	270	0	4.59 * ↑	
Moderate clutter	266	4	3.55	229.07 * ↓
High clutter	259	11	2.04	215.51 * ↓

East

S

No clutter	270	0	4.59 * ↑	
Moderate clutter	270	0	4.59 * ↑	
High clutter	105	165	73.52 * ↓	21.87 * ↓

T

No clutter	270	0	4.59 * ↑	
Moderate clutter	251	19	0.83	200.52 * ↓
High clutter	78	192	106.67 * ↓	8.54 * ↓

Table 48 continued.

V

No clutter	264	6	3.08	225.15 * ↓
Moderate clutter	240	30	0.04	180.80 * ↓
High clutter	89	181	92.42 * ↓	13.23 * ↓

X

No clutter	186	84	10.97 * ↓	98.77 * ↓
Moderate clutter	227	43	0.42	158.80 * ↓
High clutter	98	172	81.52 * ↓	17.83 * ↓

Z

No clutter	270	0	4.59 * ↑	
Moderate clutter	246	24	0.34	191.43 * ↓
High clutter	54	216	141.30 * ↓	1.86

5.17 Presence of bats at defined areas in the facility

Table 49: Chi² values of observed numbers at which bats were present at each of the nine areas in the flight room. Expected number was = 44.44 for each area. Critical value = 3.85 at d.f. = 1 for single cells. Cells with no significant test result are underlined.

Individual	wall	cable 1	cable 2	cable 3	roost	edge	ceiling	feeding area	flying
A	25.34	57.52	_0.01_	19.50	28.263	48.78	_0.44_	20.85	22.24
B	2385.00	44.44	44.44	44.44	44.440	36.80	44.44	12.36	35.00
E	783.18	19.50	16.94	44.44	35.003	62.16	4.06	36.80	44.44
H	425.80	82.53	42.70	44.44	44.440	35.00	31.54	28.26	36.80
J	44.44	_3.55_	71.99	69.46	_1.246_	4.06	44.44	_0.27_	_2.94_
K	44.44	4.06	227.55	338.01	44.440	25.16	40.53	4.06	22.24
N	44.44	11.33	62.16	646.95	38.643	8.50	33.25	20.85	14.56
O	42.46	4.06	42.70	834.37	16.943	29.88	35.00	29.88	35.00
P	44.44	22.24	12.49	1104.61	23.680	25.16	29.88	12.36	42.46
S	44.44	42.70	38.64	_4.06_	44.440	624.26	_3.55_	31.54	38.64
T	40.53	62.16	42.46	38.64	40.530	572.89	10.46	12.36	36.80
V	38.64	88.07	33.25	19.50	40.530	425.80	25.34	33.25	42.46
X	44.44	99.69	20.85	16.94	42.463	609.36	0.67	28.26	44.44
Z	40.53	131.90	22.24	22.24	44.440	310.99	19.66	25.16	36.80
1	28.26	57.52	29.88	20.85	44.440	285.10	66.98	18.20	40.53
2	44.44	38.87	44.44	31.54	40.530	523.73	44.68	26.69	28.26
3	44.44	93.79	36.80	23.68	36.800	483.34	_2.51_	29.88	16.94
4	38.64	115.23	12.36	2.01	44.440	332.51	_1.25_	31.54	19.50
5	44.44	26.88	10.34	15.73	40.530	537.55	_2.51_	16.94	31.54
6	40.53	40.76	44.44	29.88	42.463	662.30	_2.51_	12.36	26.69
7	40.53	21.02	11.33	12.36	44.440	407.43	13.57	14.56	22.24
Σ	4329.42	1067.82	868.02	3383.67	783.142	6050.81	457.29	446.47	640.56

5.18 Allogrooming events of bat dyads in the facility

Table 50: Number of allogrooming events for each dyad. Dyads of bats from the same sample site were highlighted in dark grey. First column and last row show sums of counted frequencies per bat.

	Ind	A	B	H	N	O	P	3	S	T	V	5	X	6	1
3	B	3													
6	H	4	2												
2	N	1		1											
6	O	1	1		4										
5	P				3	2									
0	3														
5	S	2				2	1								
4	T	1						1	2						
3	V								2	1					
5	5			1			1	2			1				
10	X	1				2	1	1	1	1	2	1			
8	6	1							1	3		1	2		
9	1							2	3		2	1	1		
6	7								1	1		2		2	
sums		14	3	2	7	6	3	6	10	6	5	5	3	2	

5.19 Additional descriptive results of sequence analyses

Table 51: Diagnostic sites of cyt-b sequences. Indels and 19 Polymorphic sites (100% transitions)

	30	32	60	90	109	138	184	195	207	231	270	279	285	194	306	312	343	351	371
	T	G	G	C	A	A	A	C	T	A	T	G	G	T	C	T	T	T	T
1_west'	.	.	.	.	.	.	.	.	.	.	.	.	.	.	.	.	.	.	.
13_west'	.	.	.	.	.	.	.	.	.	.	.	.	.	.	.	.	.	.	.
18_18'	.	.	.	.	.	.	.	.	.	.	.	.	.	.	.	.	.	.	.
26_west'	.	.	.	.	.	.	.	.	.	.	.	.	.	.	.	.	.	.	.
29_west'	.	.	.	.	.	.	.	.	.	.	.	.	.	.	.	.	.	.	.
1_10_central''	C	A	.	.	G	G	.	.	.	G	C	A	.	C	T	.	.	C	.
2_10_central''	C	A	.	.	G	G	.	.	.	G	C	A	.	C	T	.	.	C	.
7_10_central''	C	A	.	.	G	G	.	.	.	G	C	A	.	C	T	.	.	C	.
14_10_central''	C	A	.	.	G	G	.	.	.	G	C	A	.	C	T	.	.	C	.
16_10_central''	C	A	.	.	G	G	.	.	.	G	C	A	.	C	T	.	.	C	.
5_island'	.	.	.	.	.	.	T	.	.	.	.	.	A	.	.	.	.	.	.
28_island'	.	A	.	T	.	.	G	.	.	.	.	.	A	.	.	.	.	.	.
33_island'	.	A	.	T	.	.	.	.	.	.	.	.	A	.	.	.	.	.	.
36_island'	.	A	.	T	.	.	.	.	.	.	.	.	A	.	.	.	.	.	.
40_island'	.	A	.	T	.	.	.	.	.	.	.	.	A	.	.	.	.	.	.
9_east'	.	A	.	T	.	.	.	.	C	.	.	.	A	.	.	.	C	.	.
21_east'	.	A	.	T	.	.	.	.	C	.	.	.	A	.	.	.	C	.	.
20_east'	.	A	.	T	.	.	.	.	C	.	.	.	A	.	.	.	C	.	.
66_east'	.	A	.	T	.	.	.	.	C	.	.	.	A	.	.	.	C	.	.
37_east'	.	A	.	T	.	.	.	.	C	.	.	.	A	.	.	.	C	.	.

Amino acid changes in cyt-b. 14 Non-synonymous substitutions:
1. G (west & central) → A (island & east) in position 32;
2. G (west, island & east) → A (central) position 60;
3. A (west, central and east) → G (central) position 109;
4. A (west, central and east) → G (central) position 138;
5. A → G (one island individual) position 184;
6. C → T (one island individual) position 195;
7. A (west, island, east) → G (central) position 231;
8. A (central, island, east) → G (west) position 279;
9. A (island, east) → G (west, central) position 285;
10. C (west, island, east) → T (central) position 306;
11. T (west, island, east) → C (central) position 312;
12. T (west, central, island) → C (east) position 343; T (west, island, east) → C (central) position 351;

T → C (one central individual) position 371).

Table 52: List of haplotypes in the cyt-b data set. Above each sequence, haplotype label (A-G), frequency and sample site are given. Table continued next page.

W1;5
AAAACCCACCCACTCTTAAAAATCATTAATAGCTCATTTGTAGACCTGCCAGCACCCTCGAGCCTATCTT
CATGATGAAACTTCGGGTCCCTCCTAGGTGTATGCTTGACTGTACAAATTATAACAGGACTATTCTTAGC
TATGCACTACACGTCCGATACAACAACCGCATTCAACTCCGTAACACATTTGCCGAGATATTAATTAC
GGATGAATTCTCCGCTATCTACATGCTAATGGAGCATCCATATTCTTCATCTGCCTCTATCTTCATGTGG
GCCGGGGCCTATATTATGGATCCTACACCTATTTAGAAACATGAAATATCGGAATTATTCTCTTGTTCAC
TGTCATAGCAACTGCATTCATAGGATATGTACTACCATGAGGACAAATATCATTCTGAGGGGGCTGCA

C1;1
AAAACCCACCCACTCTTAAAAATCATTAACAGCTCATTTGTAGACCTGCCAGCACCCTCAAGCCTATCTT
CATGATGAAACTTCGGGTCCCTCCTAGGTGTATGCTTGGCTGTACAAATTATAACAGGACTATTCTTGGC
TATGCACTACACGTCCGATACAACAACCGCATTCAACTCCGTAACACACATTTGCCGAGATATTAATTAC
GGATGAATTCTCCGCTATCTGCATGCTAATGGAGCATCCATATTCTTCATCTGCCTCTACCTTCATGTAG
GCCGGGGCCTATACTATGGATCCTATACCTACTTAGAAACATGAAATATCGGAATTATTCTCTTGTTCAC
CGTCATAGCAACTGCATTCATAGGATATGTACTACCATGAGGACAAATATCATTCTGAGGGGGCTGCA

C2;1
AAAACCCACCCACTCTTAAAAATCATTAACAGCTCATTTGTAGACCTGCCAGCACCCTCAAGCCTATCTT
CATGATGAAACTTCGGGTCCCTCCTAGGTGTATGCTTGGCTGTACAAATTATAACAGGACTATTCTTGGC
TATGCACTACACGTCCGATACAACAACCGCATTCAACTCCGTAACACACATTTGCCGAGATATTAATTAC
GGATGAATTCTCCGCTATCTGCATGCTAATGGAGCATCCATATTCTTCATCTGCCTCTACCTTCATGTAG
GCCGGGGCCTATACTATGGATCCTATACCTACTTAGAAACATGAAATATCGGAATTATTCTCTTGTTCAC
CGTCATAGCAACTGCATTCACAGGATATGTACTACCATGAGGACAAATATCATTCTGAGGGGGCTGCA

C3;3
AAAACCCACCCACTCTTAAAAATCATTAACAGCTCATTTGTAGACCTGCCAGCACCCTCAAGCCTATCTT
CATGATGAAACTTCGGGTCCCTCCTAGGTGTATGCTTGGCTGTACAAATTATAACAGGACTATTCTTGGC
TATGCACTACACGTCCGATACAACAACCGCATTCAACTCCGTAACACACATTTGCCGAGATATTAATTAC
GGATGAATTCTCCGCTATCTGCATGCTAATGGAGCATCCATATTCTTCATCTGCCTCTACCTTCATGTAG
GCCGGGGCCTATATTATGGATCCTATACCTACTTAGAAACATGAAATATCGGAATTATTCTCTTGTTCAC
CGTCATAGCAACTGCATTCATAGGATATGTACTACCATGAGGACAAATATCATTCTGAGGGGGCTGCA

Table 52 continued.

I1;4
AAAACCCACCCACTCTTAAAAATCATTAATAACTCATTTGTAGACCTGCCAGCACCCTCGAGCCTATCTT
CATGATGAAACTTCGGGTCTCTCCTAGGTGTATGCTTGACTGTACAAATTATAACAGGACTATTCTTAGC
TATGCACTACACGTCCGATACAACAACCGCATTCAACTCCGTAACACACATTTGCCGAGATATTAATTAC
GGATGAATTCTCCGCTATCTACATGCTAATGGAGCATCCATATTCTTCATCTGCCTCTACCTTCATGTAG
GCCGAGGCCTATATTATGGATCCTACACCTATTTAGAAACATGAAATATCGGAATTATTCTCTTGTTCAC
TGTCATAGCAACTGCATTCATAGGATATGTACTACCATGAGGACAAATATCATTCTGAGGGGGCTGCA

I2;1
AAAACCCACCCACTCTTAAAAATCATTAATAACTCATTTGTAGACCTGCCAGCACCCTCGAGCCTATCTT
CATGATGAAACTTCGGGTCTCTCCTAGGTGTATGCTTGACTGTACAAATTATAACAGGACTATTCTTAGC
TATGCACTACACGTCCGATACAACAACCGCATTCAACTCCGTAGCACACATTTGTCGAGATATTAATTAC
GGATGAATTCTCCGCTATCTACATGCTAATGGAGCATCCATATTCTTCATCTGCCTCTACCTTCATGTAG
GCCGAGGCCTATATTATGGATCCTACACCTATTTAGAAACATGAAATATCGGAATTATTCTCTTGTTCAC
TGTCATAGCAACTGCATTCATAGGATATGTACTACCATGAGGACAAATATCATTCTGAGGGGGCTGCA

E1;5
AAAACCCACCCACTCTTAAAAATCATTAATAACTCATTTGTAGACCTGCCAGCACCCTCGAGCCTATCTT
CATGATGAAACTTCGGGTCTCTCCTAGGTGTATGCTTGACTGTACAAATTATAACAGGACTATTCTTAGC
TATGCACTACACGTCCGATACAACAACCGCATTCAACTCCGTAACACACATTTGCCGAGATATTAACTAC
GGATGAATTCTCCGCTATCTACATGCTAATGGAGCATCCATATTCTTCATCTGCCTCTACCTTCATGTAG
GCCGAGGCCTATATTATGGATCCTACACCTATTTAGAAACATGAAATATCGGAATTATTCTCCTGTTCAC
TGTCATAGCAACTGCATTCATAGGATATGTACTACCATGAGGACAAATATCATTCTGAGGGGGCTGCA

Table 53: Diagnostic sites of d-loop sequences. Indels and 14 Polymorphic sites (13 transitions, 1 transversion).

	330	327	316	306	299	298	296	293	281	270	268	262	242	230	224	203	201	195	192	191	190	183	182	179	178	170	162	159	131	96	95	88	77	71	56	54	48	44	42	41	35	33
	A	G	C	T	C	T	A	C	G	G	T	G	T	A	A	T	T	C	T	T	G	A	G	C	T	G	A	T	A	G	A	T	-	A	T	A	T	G	T	C	A	G
MW1_west'	.	.	.	.	.	.	.	.	.	.	.	.	.	.	.	.	.	.	.	.	.	.	.	.	.	.	.	.	.	.	.	.	.	.	.	.	.	.	.	.	.	.
MW13_west'	.	.	.	.	.	.	.	.	.	.	.	.	.	.	.	.	.	.	.	.	.	.	.	.	.	.	.	.	.	.	.	.	.	.	.	.	.	.	.	.	.	.
MW18_west'	.	.	.	.	.	.	.	.	.	.	.	.	.	.	.	.	.	.	.	.	.	.	.	.	.	.	.	.	.	.	.	.	.	.	.	.	.	.	.	.	.	.
MW26_west'	.	.	.	.	.	.	.	.	.	.	.	.	.	.	.	.	.	.	.	.	.	.	.	.	.	.	.	.	.	.	.	.	.	.	.	.	.	.	.	.	.	.
MW29_west'	.	.	.	.	.	.	.	.	.	.	.	.	.	.	.	.	.	.	.	.	.	.	.	.	.	.	.	.	.	.	.	.	.	.	.	.	.	.	.	.	.	.
MW5_island'	G	.	T	.	T	C	G	T	.	.	.	.	.	G	G	.	.	.	C	.	.	G	.	T	.	A	G	.	.	.	.	C	.	G	.	.	.	.	.	.	.	A
MW28_island'	G	.	T	.	T	C	G	T	.	A	.	.	.	G	G	.	.	.	C	.	.	G	.	T	.	A	G	.	.	.	.	C	.	.	.	.	.	.	.	.	.	A
MW33_island'	G	.	T	.	T	C	G	T	.	.	C	.	.	G	G	C	.	.	C	.	.	G	.	T	.	A	G	.	.	.	.	C	.	G	.	.	.	.	.	.	.	A
MW38_island'	G	.	T	.	T	C	G	T	.	.	.	.	.	G	G	.	.	.	C	.	.	G	.	T	.	A	G	.	.	.	.	C	.	G	.	.	.	.	.	.	.	A
MW40_island'	G	.	T	.	T	C	G	T	.	.	.	.	.	G	G	C	.	.	C	.	.	G	.	T	.	A	G	.	.	.	.	C	.	G	.	.	.	.	.	.	.	A
MW1_10_ce'	G	.	.	.	.	.	G	.	.	.	.	.	C	.	.	.	.	T	.	C	A	.	.	.	.	.	.	.	.	.	.	.	.	.	G	.	.	.	.	.	.	.
MW2_10_ce'	G	.	.	.	T	.	G	.	.	.	.	.	C	.	.	.	.	T	.	C	A	.	.	.	.	.	.	.	.	.	.	.	.	.	G	.	.	.	.	.	.	.
MW7_10_ce'	G	.	.	.	T	.	G	.	.	.	.	.	C	.	.	.	.	T	.	C	A	.	.	.	.	.	.	.	.	A	.	.	.	.	G	.	.	.	.	.	.	.
MW14_10_c'	G	.	.	.	T	.	G	.	.	.	.	.	C	.	.	.	.	T	.	C	A	.	.	.	.	.	.	.	.	A	.	.	.	.	G	.	.	.	.	.	.	.
MW16_10_c'	G	.	.	.	.	.	G	.	.	.	.	.	C	.	.	.	.	T	.	C	A	.	.	.	.	.	.	.	.	A	G	.	.	.	G	C	A	.	.	.	G	.
MW9_east'	G	.	.	.	.	C	G	.	A	.	.	.	.	G	G	.	.	.	C	C	A	G	A	.	.	.	.	.	G	.	.	C	.	.	.	G	.	A	.	.	G	.
MW20_east'	G	T	.	.	.	C	G	.	A	.	.	.	.	G	G	.	.	.	C	C	A	G	A	.	.	.	.	.	G	.	.	C	.	.	.	G	.	A	.	.	G	.
MW21_east'	G	.	.	.	.	C	G	.	A	.	.	.	.	.	G	.	.	.	C	C	A	G	.	.	.	.	A	.	G	.	.	C	.	.	.	G	.	A	.	.	G	.
MW37_east'	G	.	.	.	.	C	G	.	A	.	.	.	.	G	G	.	.	.	C	C	A	G	.	.	.	.	.	C	G	.	.	C	.	.	.	.	.	A	.	.	G	.
MW66_east'	G	T	.	.	C	C	G	.	A	.	.	A	.	G	G	.	C	.	C	C	A	G	A	.	.	.	.	.	G	.	.	.	.	G	G	G	.	A	.	.	G	.

A. Bastian *Cuban Macrotus variability*

23 Amino acid changes:
1. K (west, island, central) R → (east) in position 12
2. P (west, central, island & east) → L (1 island) in position 14
3. C (west, central & island) → Y (east) in position 15
4. I (west, central island & east) → T (2 island) in position 19
5. - (west, central & east) → G (4 island) & D (1 island) in position 24
6. L → P (5 island; 1east) in position 26
7. T → A (1 east) in position 30
8. R → K (3 central) in position 32
9. L → S (1 east) in position 44
10. S → L (5 island) in position 57
11. E → G (5 island; 1 east) & R (4 east) in position 60
12. C → Y (3 central; 5 east) in position 61
13. P → S (5west) & F (4 central) in position 64
14. N → S (5 island; 1 east) in position 68
15. F → S (5 central) in position 75
16. G → E (1east) in position in position 77
17. V → I (1 island) in position 88
18. V → I (1 central; 5 east) in position 90
19. C → Y (2west) in position 94
20. I → T (5 island; 5 east) in position 98
21. T → I (1 island; 3central) in position 99
22. S → L (5 island; 2 east) in position 100
23. G → S (5west; 1 island; 4 east) in position 106

Table 54: List of haplotypes in the d-loop data set. Above each sequence, haplotype label (A-G), frequency and sample site are given. Table continued next pages.

<u>W1;2</u>
TGTTGCTGGTTTCACGGAGGTAGCCAAATCAAGAAATTGCCTTGTACTGAGTCAATAGTCATGAATTTGG
ACGACTGAATGTGATGACTAATAGAATGGTGTATGCACGATTACTAAATTGTATGTACTTAGATCATTAT
ATTATTGTCCTATAATAATAGTATATGTCAGATTATGAATGTTTATGTTCTGTTCCATTAAATTAATTTA
CTTGCTTAATATTTCATGGGGTACATAATTTAATGTAATATTATACATAGTATGTGTCTATAACATATAT
ATGGGGTATATAACATCATGCACGAAGTACATAAGCTTTTTTGCGGGATCAAGGAGTAGTTTAAGTAGAA
CTTCAGCTTTGGGTGTTGAGGGTGGGG

<u>W2;2</u>
TGTTGCTGGTTTCACGGAGGTAGCCAAATCAAGAAATTGCCTTGTACTGAGTCAATAGTCATGAATTTGG
ACGACTGAATGTGATGACTAATAGAATGGTGTATGCACGATTACTAAATTGTATGTACTTAGATCATTAT
ATTATTGTCCTATAATAATAGTATATGTCAGATTATGAATGTTTATGTTCTGTTCCATTAAATTAATTTA
CTTGCTTAATATTTCATGGGGTACATAATTTAATGTAATATTATACATAGTATGTGTCTATAACATATGT
ATGGGGTATATAACATCATGCACGAAGTACATAAGCTTTTTTGCGGGATCAAGGAGTAGTTTAAGTAGAA
CTTCAGCTTTGGGTGTTGAGGGTGGGG

<u>W3;1</u>
TGTTGCTGGTTTCACGGAGGTAGCCAAATCAAGAAATTGCCTTGTACTGAGTCAATAGTCATGAATTTGG
ACGACTGAATGTGATGACTAATAGAATGGTGTATGCACGATTACTAAATTGTATGTACTTAGATCATTAT
ATTATTGTCCTATAATAATAGTATATGTCAGATTATGAATGTTTATGTTCTGTTCCATTAAATTAATTTA
CTTGCTTAATATTTCATGGGGTACATAATTTAATGTAATATTATACATAGTATGTGTCTATAACATATGT
ATGGGGTATATAACATCATGCACGAAGTACATAAGCTTTTTGCGGGATCAAGGAGTAGTTTAAGTAGAAC
TTCAGCTTTGGGTGTTGAGGGTGGGG

Table 54 continued.

C1;1
```
TGTTGCTGGTTTCACGGAGGTAGCCAAATCAAGAAATTGCCTTGTACTGAGTCGATAGTCATGAATTTGG
ACGACTGAATGTGATGACTAATAGAATGGTGTATGCACGATTACTAAATTGTATGTACTTAGATCATTAT
ATTATTGTCCTATAATAATAGATATGTCAGATTATGAATACTTATGTTTTGTTCCATTAAATTAATTTAC
TTGCTTAATATCTCATGGGGTACATAATTTAATGTAATATTATACATAGTATGTGTTTATAACATATGTA
TGGGGTATATAATATCATGCACGAAGTACATAGGCTTTTTTGCGGGATCAAGGAGTAGTTTAAGTAGAAC
TTCAGCTTTGGGTGTTGAGGGTGGGG
```

C2;1
```
TGTTGCTGGTTTCACGGAGGTAGCCAAATCAAGAAATTGCCTTGTACTGAGTCGATAGTCATGAATTTGG
ACGACTGAATGTGATGACTAATAGAATGGTGTATGCACGATTACTAAATTGTATGTACTTAGATCATTAT
ATTATTGTCCTATAATAATAGATATGTCAGATTATGAATACTTATGTTTTGTTCCATTAAATTAATTTAC
TTGCTTAATATCTCATGGGGTACATAATTTAATGTAATATTATACATAGTATGTGTCTATAACATATGTA
TGGGGTATATAATATCATGCACGAAGTACATAGGCTTTTTTGCGGGATCAAGGAGTAGTTTAAGTAGAAC
TTCAGCTTTGGGTGTTGAGGGTGGGG
```

C3;1
```
TGTTGCTGGTTTCACGGAGGTAGCCAAATCAAGAAATTGCCTTGTACTGAGTCGATAGTCATGAATTTGG
ACGACTGAATGTGATGACTAATAAAATGGTGTATGCACGATTACTAAATTGTATGTACTTAGATCATTAT
ATTATTGTCCTATAATAATAGTATATGTCAGATTATGAATACTTATGTTTTGTTCCATTAAATTAATTTA
CTTGCTTAATATCTCATGGGGTACATAATTTAATGTAATATTATACATAGTATGTGTCTATAACATATGT
ATGGGGTATATAATATCATGCACGAAGTACATAGGCTTTTTTGCGGGATCAAGGAGTAGTTTAAGTAGAA
CTTCAGCTTTGGGTGTTGAGGGTGGGG
```

C4;1
```
TGTTGCTGGTTTCACGGAGGTAGCCAAATCAAGAAATTGCCTTGTACTGAGTCGATAGTCATGAATTTGG
ACGACTGAATGTGATGACTAATAAAATGGTGTATGCACGATTACTAAATTGTATGTACTTAGATCATTAT
ATTATTGTCCTATAATAATAGTATATGTCAGATTATGAATACTTATGTCTTGTTCCATTAAATTAATTTA
CTTGCTTAATATCTCATGGGGTACATAATTTAATGTAATATTATACATAGTATGTGTCTATAACATATGT
ATGGGGTATATAACATCATGCACGAAGTACATAGGCTTTTTGCGGGATCAAGGAGTAGTTTAAGTAGAAC
TTCAGCTTTGGGTGTTGAGGGTGGGG
```

C5;1
```
TGTTGCTGGTTTCACGGAGGTAGCCAAATCAAGAAATTGCCTTGTACTGAGTCGATAGTCATGAATTTGG
ACGACTGATGTGATGACTAATAAGATGGTGTATGCACGATTACTAAATTGTATGTACTTAGATCATTAT
ATTATTGTCCTATAATAATAGTATATGTCAGATTATGAATGTTTATGTTTTGTTCCATTAAATTAATTTA
CTTGCTTAATATCTCATGGGGTACATAATTTAATGTAATATTATACATAGTATGTATCTATAACATATGT
ATGGGGTATATAACATCATGCACGAAGTACATAGGCTTTTTTGCGGGATCAAGGAGTAGTTTAAGTAGAA
CTTCAGCTTTGGGTGTTGAGGGTGGGG
```

I1;1
```
TGTTGCTGGTTTCACGGAGGTAGCCAAATCAAAAAATTGCCTTGTACTGAGTCAATAGTCATGAATTTGG
GTCGACCGAATGTGATGACTAATAGAATGGTGTATGCACGATTACTAAATTGTATGTACTTAGATCATTA
TATTATTGTCCTATAATAGTAATATGTTAGATTATGGATGTTTATGTCCTGTTCCATTGAGTTAATTT
ACTTGCTTAATATTTCATGGGGTACATAATTTAATGTAATATTATACATAGTATGTGTTTATAACATATG
TATGGGGTATACAACATTATGCACGAAGTACATAGGCTTTTTTGCGGGATCAAGGAGTAGTTTAAGTAGA
ACTTCAGCTTTGGGTGTTGAGGGTGGGG
```

Table 54 continued.

I2;1
TGTTGCTGGTTTCACGGAGGTAGCCAAATCAAAAAATTGCTCTGTACTGAGTCAATAGTCATGAATTTGG
ATCGACCGAATGTGATGACTAATAGAATGGTGTATGCACGATTACTAAATTGTATGTACTTAGATCATTA
TATTATTGTCCTATAATAGTAATATATGTTAGATTATGGATGTTTATGTCCTGTTCCATTGAGTTAATTT
ACTTGCTTAATATTTCATGGGGTACATAATTTAATGTAATATTATACATAATATGTGTTTATAACATATG
TATGGGGTATACAATACTATGCACGAAGTACATAAGCTTTTTGCGGGATCAAGGAGTAGTTTAAGTAGA
ACTTCAGCTTTGGGTGTTGAGGGTGGGG

I3;2
TGTTGCTGGTTTCACGGAGGTAGCCAAATCAAAAAATTGCCTTGTACTGAGTCAACAGTCATGAATTTGG
GTCGACCGAATGTGATGACTAATAGAATGGTGTATGCACGATTACTAAATTGTATGTACTTAGATCATTA
TATTATTGTCCTATAATAGTAATATATGTTAGATTATGGATGTTTATGTCCTGTCCCATTGAGTTAATTT
ACTTGCTTAATATTTCATGGGGTACATAATTTAATGTAATATTATACATAGTATGTGTTTATAACATATG
TATGGGGTATACAACATTATGCACGAAGTACATAGGCTTTTTGCGGGATCAAGGAGTAGTTTAAGTAGAA
CTTCAGCTTTGGGTGTTGAGGGTGGGG

I4;1
TGTTGCTGGTTTCACGGAGGTAGCCAAATCAAAAAATTGCCTTGTACTGAGTCAATAGTCATGAATTTGG
GTCGACCGAATGTGATGACTAATAGAATGGTGTATGCACGATTACTAAATTGTATGTACTTAGATCATTA
TATTATTGTCCTATAATAGTAATATATGTTAGATTATGGATGTTTATGTCCTGTTCCATTGAGTTAATTT
ACTTGCTTAATATTTCATGGGGTACATAATCTAATGTAATATTATACATAGTATGTGTCTATAACATATG
TATGGGGTATACAACATTATGCACGAAGTACATAGGCTTTTTGCGGGATCAAGGAGTAGTTTAAGTAGA
ACTTCAGCTTTGGGTGTTGAGGGTGGGG

E1;1
TGTTGCTGGTTTCACGGAGGTAGCCAAATCAAGAGATTGCCTTATACCGAGTCGATAGTCATGAATTTGG
ACGACCGAATGTGATGGCTAATAGGATGGTGTATGCACGATTACTAAATTGTATGTACTTAGATCATTAT
ATTATTGTCCTATAATAATAGTATATGTCAGATTATAGATATTTATGTCCTGTTCCATTGAGTTAATTTA
CTTGCTTAATATTTCATGGGGTACATAATTTAATGTAATATTATACATAGTATGTATCTATAACATATGT
ATGGGGTATACAACATTATGCACGAAGTACATAGGCTTTTTGCAGGGTCAAGGAGTAGTTTAAGTAGAAC
TTCAGCTTTGGGTGTTGAGGGTGGGG

E2;2
TGTTGCTGGTTTCACGGAGGTAGCCAAATCAAGAGATTGCCTTATACTGAGTCGATAGTCATGAATTTGG
ACGACTGAATGTGATGACTAATAGGATGGTGTATGCACGATTACTAAATTGTATGTACTTAGATCATTAT
ATTATTGTCCTATAATAATAGTATATGTCAGATTATAGATACTTATGTCCTGTTCCATTGAATTAATTTA
CTTGCTTAATATTTCATGGGGTACATAATTTAATGTAATATTATACATAGTATGTATCTATAACATATGT
ATGGGGTATACAACATCATGCACGAAGTACATAAGCTTTTTGCAGGGTCAAGGAGTAGTTTAAGTAGAAC
TTCAGCTTTGGGTGTTGAGGGTGGGG

E3;1
TGTTGCTGGTTTCACGGAGGTAGCCAAATCAAGAGATTGCCTTATACTGAGTCAATAGTCATGAATTTGG
ACGACTGAATGTGATGACTAATAGGATGGTGTATGCACGATTACTAAATTGTATGTACTCAGATCATTAT
ATTATTGTCCTATAATAATATATGTCAGATTATGGATATTTATGTCCTGTTCCATTGAATTAATTTA
CTTGCTTAATATTTCATGAGGTACATAATTTAATGTAATATTATACATAGTATGTATCTATAACATATGT
ATGGGGTATACAACATTATGCACTAAGTACATAAGCTTTTTGCAGGGTCAAGGAGTAGTTTAAGTAGAAC
TTCAGCTTTGGGTGTTGAGGGTGGGG

E4;1
TGTTGCTGGTTTCACGGAGGTAGCCAAATCAAGAGATTGCCTTATACTGAGTCGATAGTCATGAATTTGG
ACGACTGAATGTGATGACTAATAGGATGGTGTATGCACGATTACTAAATTGTATGTACTTAGATCATTAT
ATTATTGTCCTATAATAATAGTATATGTCAGATTATAGATACTTATGTCCCGTTCCATTGAATTAATTTA
CTTGCTTAATATTTCATGGGGTACATAATTTAATGTAATATTATACATAGTATGTATCTATAACATATGT
ATGGGGTATACAACATCATGCACGAAGTACATAAGCTTTTTGCAGGGTCAAGGAGTAGTTTAAGTAGAAC
TTCAGCTTTGGGTGTTGAGGGTGGGG

6. List of references

Ahlén, I. (1990) *Identification of bats in flight*. Swedish Society for Nature Conservation, Stockholm.

Anderson, S. (1969) Macrotus waterhousii. *Mammalian Species*, **1**, 1-4

Anderson, S. & Nelson, C.E. (1965) A systematic revision of Macrotus (Chiroptera). *American Museum Novitates*, **2212**, 1-39

Avila-Flores, R. & Meddelín, R.A. (2004) Ecological, taxonomic, and physiological correlates of cave use by Mexican bats. *Journal of Mammalogy*, **85**, 675-687

Baker, C.J. & Genoways, H.H. (1978) Zoography of Antillean bats. *Zoography in the Caribbean* (ed. by F.B. Gill), pp. 53-97. Academy of Natural Sciences Philadelphia, Philadelphia. Pennsylvania. USA.

Baker, R.J. & Bickham, J.W. (1980) Karyotypic evolution in bats: Evidence of extensive and conservative chromosomal evolution in closely related taxa. *Systematic Zoology*, **29**, 239-252

Balcombe, J.P. & McCracken, G.F. (1992) Vocal recognition in Mexican free-tailed bats: Do pups recognize mothers? *Animal Behaviour* **43**, 79-87

Band, A.-L. (2000) Bats of the Cayman Islands. *BATS Magazine*, **18**, 9-12

Barclay, R.M.R. (1983) Echolocation calls of Emballonurid bats from Panama. *Journal of Comparative Physiology A: Neuroethology, Sensory, Neural, and Behavioral Physiology*, **151**, 515-520

Barclay, R.M.R. & Brigham, R.M. (1991) Prey detection, dietary niche breadth, and body size in bats: Why are aerial insectivorous bats so small? *The American Naturalist*, **137**, 693-703

Barclay, R.M.R., Fenton, M.B. & Thomas, D.W. (1979) Social behavior of the Little brown bat, Myotis lucifugus ii: Vocal communication. *Behavioral Ecology and Sociobiology*, **6**, 137-146

Behr, O. & Von Helversen, O. (2004) Bat serenades-complex courtship songs of the Sac-winged bat (Saccopteryx bilineata). *Behavioral Ecology and Sociobiology*, **56**, 106-115

Bell, G.P. (1985) The sensory basis of prey location by the California leaf-nosed bat Macrotus californicus (Chiroptera: Phyllostomidae). *Behavioral Ecology and Sociobiology*, **16**, 343-347

Bell, G.P. & Fenton, M.B. (1986) Visual acuity, sensitivity and binocularity in a gleaning insectivorous bat, Macrotus californicus (Chiroptera: Phyllostomidae). *Animal Behaviour*, **34**, 409-414

Bell, G.P., Bartholomew, G.A. & Nagy, K.A. (1986) The roles of energetics, water economy, foraging behavior, and geothermal refugia in the distribution of the bat, Macrotus californicus. *Journal of Comparative Physiology B*, **156**, 441-445

Berry, R.D. & Brown, P.E. (1995) Natural history and reproductive behavior of the California leaf-nosed bat (Macrotus californicus). *Bat Research News*, **36**, 49-50

Bogdanowicz, W., Fenton, M.B. & Daleszczyk, K. (1999) The relationships between echolocation calls, morphology and diet in insectivorous bats. *Journal of Zoology*, **247**, 381-393

Boonman, A. & Ostwald, J. (2007) A modeling approach to explain pulse design in bats. *Biological Cybernetics*, **97**, 159-172

Boughman, J.W. & Wilkinson, G.S. (1998) Greater spear-nosed bats discriminate group mates by vocalizations. *Animal Behaviour*, **55**, 1717-1732

Buden, D.W. (1975) A taxonomic and zoogeographic appraisal of the Big-eared bat (Macrotus waterhousii Gray) in the West Indies. *Journal of Mammalogy*, **56**, 758-769

Conservation International, F. (2005) Caribbean Islands. URL http://www.conservation.org/where/priority_areas/hotspots/north_central_america/Caribbean-Islands/Pages/biodiversity.aspx

Cox, C.B. & Moore, P.D. (2000) The source of novelty. *Biogeography an ecological and evolutionary approach* (ed. by C.B. Cox and P.D. Moore). Blackwell Science, London.

Dávalos, L.M. (2004) Phylogeny and biogeography of Caribbean mammals. *Biological Journal of the Linnean Society*, **81**, 373-394

Dávalos, L.M. & Eriksson, R. (2003) New and noteworthy records from ten Jamaican bat caves. *Caribbean Journal of Science*, **39**, 140-144

Dávalos, L.M. & Jansa, S.A. (2004) Phylogeny of the Lonchophyllini (Chiroptera: Phyllostomidae). *Journal of Mammalogy*, **85**, 404-413

Davidson, S.M. & Wilkinson, G.S. (2004) Function of male song in the Greater white-lined bat, Saccopteryx bilineata. *Animal Behaviour*, **67**, 883-891

Davis, B.L. & Baker, R.J. (1974) Morphometrics, evolution, and cytotaxonomy of mainland bats of the genus Macrotus (Chiroptera: Phyllostomatidae). *Systematic Zoology*, **23**, 26-39

Dobson, G.E. (1878) Macrotus. *Catalogue of the Chiroptera in the collection of the British Museum*, 463-467

Doerrie, M., Schmidt, S. & Suba, M.S.K. (2001) Contact calls of the bat, Megaderma lyra: A comparison between an Indian and a Sri Lankan population. *Zoology*, **104(Suppl IV)**, 5

Esser, K.-H. & Schmidt, U. (1989) Mother infant communication in the Lesser spear-nosed bat Phyllostomus discolor (Chiroptera, phyllostomidae) - evidence for acoustic learning. *Ethology*, **82**, 156-168

Excoffier, L., Smouse, P.E. & Quattro, J.M. (1992) Analysis of molecular variance inferred from metric distances among DNA haplotypes: Application to human mitochondrial DNA restriction data. *Genetics*, **131**, 479-491

Fenton, M.B. (1990) The foraging behaviour and ecology of animal-eating bats. *Canadian Journal of Zoology*, **68**, 411-422

Fenton, M.B. & Bell, G.P. (1981) Recognition of species of insectivorous bats by their echolocation calls. *Journal of Mammalogy*, **62**, 233-243

Fenton, M.B. & Ratcliffe, J. (2004) Eavesdropping on bats. *Nature*, **429**, 612-613

Fleming, T.H. (1992) How do fruit- and nectar-feeding birds and mammals track their food resources ? *Effects of resource distribution on plant-animal interactions* (ed. by M.D. Hunter, T. Ohgushi and P.W. Price), pp. 355-391. Academic Press, Orlando, Florida.

Fleming, T.H. & Murray, K.L. (2009) Population and genetic consequences of hurricanes for three species of West Indian phyllostomid bats. *Biotropica*, **41**, 250-256

Fleming, T.H., Murray, K.L. & Carstens, B. (2009) Phylogeography and genetic structure of three evolutionary linneages of West Indian phyllostomid bats. *Island bats: Evolution, ecology and conservation* (ed. by T.H. Fleming and P.A. Racey), pp. 116-150. The University of Chicago Press, Chicago.

Flower, W.H. & Lydekker, R. (1891) *An introduction to the study of mammals living and extinct.* Adam and Charles Black, London.

Freeman, P.W. (2000) Macroevolution in microchiroptera: Recoupling morphology and ecology with phylogeny. *Evolutionary Ecology Research,* **2**, 317-334

Gelfand, D.L. & McCracken, G.F. (1986) Individual variation in the isolation calls of Mexican free-tailed bat pups (Tadarida brasiliensis mexicana). *Animal Behaviour,* **34**, 1078-1086

Genaro, J.A. (2006) A history of systematic studies of the bees of Cuba (Insecta: Hymenoptera, Anthophila. *Zootaxa,* **1195**, 39-60

Genaro, J.A. & Tejuca, A.E. (1999) Datos cuantitativos, endemismo y estado actual del conocimiento de los insectos cubanos. *Cocuyo,* **8**, 24-28

Genoways, H.H., Baker, R.J., Bickham, J.W. & Phillips, C.J. (2005) Bats of Jamaica. *Special Publications, The Museum, TexasTech University,* **48**, 1-155

Gould, E. (1975) Neonatal vocalizations in bats of eight genera. *Journal of Mammalogy,* **56**, 15-19

Gould, E., Woolf, N.K. & Turner, D.C. (1973) Double-note communication calls in bats: Occurrence in three families. *Journal of Mammalogy,* **54**, 998-1001

Grant, B.R. & Grant, P.R. (2002) Simulating secondary contact in allopatric speciation: An empirical test of premating isolation. *Biological Journal of the Linnean Society,* **76**, 545-556

Gray, J.E. (1843) Macrotus. *Proceedings of the Zoological Society of London,* 20-21

Greenbaum, I.F. & Baker, R.J. (1976) Evolutionary relationships in Macrotus (Mammalia: Chiroptera): Biochemical variation and karyology. *Systematic Zoology,* **25**, 15-25

Griffith, T.A. & Klingener, D. (1988) On the distribution of greater Antillean bats. *Biotropica,* **20**, 240-251

Gundlach, J. (1864) Sitzung der physikalisch-mathematischen Klasse. *Monatsberichte der Königlichen Preußischen Akademie der Wissenschaften zu Berlin,* 382-383

Gundlach, J. (1872) Catálogo de los mamíferos cubanos. *Anales de la Sociedad Espanola de Historia Natural,* **40**, 239

Hasegawa, M., Kishino, H. & Yano, T.-A. (1985) Dating of the human-ape splitting by a molecular clock of mitochondrial DNA. *Journal of Molecular Evolution,* **22**, 160-174

Hedges, S.B. (2002) Morphological variation and the defintion of species in the snake genus Tropidophis (Serpentes, Tropidophiidae). *Bulletin of the Natural History Museum of London,* **68**, 83-89

Heller, K.-G. & Helversen, O.V. (1989) Resource partitioning of sonar frequency bands in rhinolophoid bats. *Oecologia,* **80**, 178-186

Hoffmann, F.G., Hoofer, S.R. & Baker, R.J. (2008) Molecular dating of the diversification of Phyllostominae bats based on nuclear and mitochondrial DNA sequences. *Molecular Phylogenetics and Evolution,* **49**, 653-658

Howell, N., Kubacka, I. & Mackey, D.A. (1996) How rapidly does the human mitochondrial genome evolve? *Journal Name: American Journal of Human Genetics; Journal Volume: 59; Journal Issue: 3; Other Information: PBD: Sep 1996,* Medium: X; Size: pp. 501-509

Itturalde-Vinent, M.A. & Macphee, R.D.E. (1999) Paleogeography of the Caribbean region: Implications for cenozoic biogeography. *Bulletin American Museum of Natural History,* **238**, 1-95

Jacobs, D.S., Barclay, R.M.R. & Walker, M. (2007) The allometry of echolocation call frequencies of insectivorous bats: Why do some species deviate from the pattern? *Oecologia*, **152**, 583-594

Jones, G. (1996) Does echolocation constrain the evolution of body size in bats? *Symposia of the Zoological Society of London*, **69**, 111-128

Jones, G. (1997a) Acoustic signals and speciation: The roles of natural and sexual selection in the evolution of cryptic species. *Advances in the Study of Behavior*, **26**, 317-354

Jones, G. (1997b) Acoustic signals and speciation: The roles of natural selection and sexual selection in the evolution of cryptic species. *Advances in the Study of Behavior*, **26**, 317-254

Jones, G. & Barlow, K.E. (2004) Cryptic species of echolocating bats. *Echolocation in bats and dolphins* (ed. by J. Thomas, C. Moss and M. Vater), pp. 345-349. The University of Chicago Press, Chicago.

Jones, K.E., Purvis, A., Maclarnon, A., Binida-Emonds, O.R.P. & Simmons, N.B. (2002) A phylogenetic supertree of the bats (Mammalia: Chiroptera). *Biological Review*, **77**, 223-259

Kiefer, A., Mayer, F., Kosuch, J., Von Helversen, O. & Veith, M. (2002) Conflicting molecular phylogenies of European long-eared bats (Plecotus) can be explained by cryptic diversity. *Molecular Phylogenetics and Evolution*, **25**, 557-566

Kingston, T., Lara, M.C., Jones, G., Akbar, Z., Kunz, T.H. & Schneider, C.J. (2001) Acoustic divergence in two cryptic Hipposideros species: A role for social selection? *Proceedings of the Royal Society of London. Series B: Biological Sciences*, **268**, 1381-1386

Kocher, T.D., Thomas, W.K., Meyer, A., Edwards, S.V., Pääbo, S., Villablanca, F.X. & Wilson, A.C. (1989b) Dynamics of mitochondrial DNA evolution in animals: Amplification and sequencing with conserved primers. *Proceedings of the National Academy of Sciences of the United States of America*, **86**, 6196-6200

Koopman, K.F. (1989) A review and analysis of the bats of the West Indies. *Biogeography of the West Indies: Past, present and future* (ed. by C.A. Woods), pp. 635-644. Sandhill Crane Press, Gainesville, Florida.

Koopman, K.F. (1994) Chiroptera: Systematics. *Handbook of zoology*. Niethammer, J., Schliemann, H., Starck, D.

Koopman, K.F., Hecht, M.K. & Ledecky-Janecek, E. (1957) Notes on the mammals of the Bahamas with special reference to the bats. *Journal of Mammalogy*, **38**, 164-173

Kulzer, E. (2005) Chiroptera. Handbuch der Zoologie (Handbook of zoology). In. Walter de Gruyter, Berlin

Leblanc, D. (2001) Adoption of a Lesser short-nosed fruit bat (Cynopterus brachyotis) in captivity. *Animal Keeper´s Forum*, **28**, 444-447

Leippert, D. (1994) Social behavior on the wing in the False vampire, Megaderma lyra. *Ethology*, **98**, 111-127

Leippert, D., W., G., H., H. & J., M.G.B. (2000) Roost-mate communication in adult Indian False vampire bats (Megaderma lyra): An indication of individuality in temporal and spectral pattern. *Animal Cognition*, **3**, 99-106

Lundberg, K. & Gerell, R. (1986) Territorial advertisement and mate attraction in the bat Pipistrellus pipistrellus. *Ethology*, **71**, 115-124

Ma, J., Kobayasi, K., Zhang, S. & Metzner, W. (2006) Vocal communication in adult Greater horseshoe bats, Rhinolophus ferrumequinum. *Journal of Comparative Physiology A*, **192**, 535-550

Macías, S., Mora, E.C. & García, A. (2006) Acoustic identification of mormoopid bats: A survey during the evening exodus. *Journal of Mammalogy*, **87**, 324-330

Mayer, F. & Von Helversen, O. (2001) Cryptic diversity in European bats. *Proceedings of the Royal Society of London B*, **268**, 1825-1831

Morgan, G.S. (2001) Patterns of extinction in West Indian bats. *Biogeography of the West Indies: Patterns and perspectives* (ed. by C.A. Woods and F.E. Sergile), pp. 369–424. CRC Press LLC, Boca Raton.

Moss, C.F. & Schnitzler, H.U. (1989) Accuracy of target ranging in echolocating bats: Acoustic information processing. *Journal of Comparative Physiology A: Neuroethology, Sensory, Neural, and Behavioral Physiology*, **165**, 383-393

Murray, K.L., Fleming, T.H., Gaines, M.S. & Williams, D.A. (2008) Characterization of polymorphic microsatellite loci for two species of phyllostomid bats from the Greater Antilles (Erophylla sezekorni and Macrotus waterhousii). *Molecular Ecology Resources*, **8**, 596-598

Murray, K.L., Fraser, E., Davy, C., Fleming, T.H. & Fenton, M.B. (2009) Characterization of the echolocation calls of bats from Exuma, Bahamas. *Acta Chiropterologica*, **11**, 415-424

Muscarella, R.A., Murray, K.L., Ortt, D., Russell, A.L. & Fleming, T.H. (2011) Exploring demographic, physical, and historical explanations for the genetic structure of two lineages of Greater Antillean bats. *PLoS ONE*, **6**, e17704

Myers, N., Mittermeier, R.A., Mittermeier, C.G., Da Fronseca, G.A.B. & Kent, J. (2000) Biodiversity hotspots for conservation priorities. *Nature*, **403**, 853-858

Naranjo, C., Muñoz Riviaux, S., Moreira, F.F.F. & Correa Court, R.C. (2010) Taxonomy and distribution of aquatic and semiaquatic Heteroptera (Insecta) from Cuba. *International Journal of Tropical Biology*, **58**, 897-907

Nelson-Rees, W.A., Kniazeff, A.J., Baker, R.J. & Patton, J.L. (1968) Intraspecific chromosome variation in the bat, Macrotus waterhousii Gray. *Journal of Mammalogy*, **49**, 706-711

Norberg, U.M. & Rayner, J.M.V. (1987) Ecological morphology and flight in bats (Mammalia; Chiroptera): Wing adaptations, flight performance, foraging strategy and echolocation. *Philosophical Transactions of the Royal Society of London. B, Biological Sciences*, **316**, 335-427

Novick, A. (1963) Orientation in Neotropical bats. II. Phyllostomatidae and Desmodontidae. *Journal of Mammalogy*, **44**, 44-55

Novick, A. (1977) *Acoustic orientation*. Academic Press, New York.

O'Farrell, M.J. & Gannon, W.L. (1999) A comparison of acoustic versus capture techniques for the inventory of bats. *Journal of Mammalogy*, **80**, 24-30

Page, R.D.M. (1996) Treeview: An application to display phylogenetic trees on personal computers. *Computer Applications in the Biosciences*, **12**, 357-358

Peck, S.B. (2005) A checklist of the beetles of Cuba with data on distributions and bionomics (Insecta: Coleoptera). In: *Arthropods of Florida and Neighboring Land Areas*, pp. 1-241. Florida Department of Agriculture and Consumer Services, Gainesville, FL

Pfalzer, G. & Kusch, J. (2003) Structure and variability of bat social calls: Implications for specificity and individual recognition. *Journal of The Zoological Society of London*, **261**, 21-33

Posada, D. & Crandall, K.A. (1998) Modeltest: Testing the model of DNA substitution. *Bioinformatics*, **14**, 817-818

Pumo, D.E., Goldin, E.Z., Elliot, B., Phillips, C.J. & Genoways, H.H. (1988) Mitochondrial DNA polymorphism in three Antillean Island populations of the Fruit bat, Artibeus jamaicensis. *Molecular Biology and Evolution*, **5**, 79-89

Rehn, J.A.G. (1904) A revision of the mammalian genus Macrotus. *Proceedings of the Academy of Natural Sciences of Philadelphia*, **56**, 427-446

Rehorek, S.J., Smith, T.D. & Bhatnagar, K.P. (2010) The orbitofacial glands of bats: An investigation of the potential correlation of gland structure with social organization. *The Anatomical Record: Advances in Integrative Anatomy and Evolutionary Biology*, **293**, 1433-1448

Ricklefs, R. & Bermingham, E. (2008) The West Indies as a laboratory of biogeography and evolution. *Philosophical Transactions of the Royal Society B*, **363**, 2393-2413

Rodríguez, F., Oliver, J.L., Marín, A. & Medina, J.R. (1990) The general stochastic model of nucleotide substitution. *Journal of Theoretical Biology*, **142**, 485-501

Rousset, F. (1997) Genetic differentiation and estimation of gene flow from F-statistics under isolation by distance. *Genetics*, **145**, 1219-1228

Russo, D. & Jones, G. (1999) The social calls of Kuhl's pipistelles pipistrellus kuhlii (Kuhl, 1819): Structure and variation. *Journal of Zoology*, **249**, 476-481

Russo, D. & Jones, G. (2000) The two cryptic species of Pipistrellus pipistrellus (Chiroptera: Verspertilionidae) occur in Italy: Evidence from echolocation and social calls. *Mammalia*, **64**, 187-196

Russo, D. & Jones, G. (2002) Identification of twenty-two bat species (Mammalia: Chiroptera) from Italy by analysis of time-expanded recordings of echolocation calls. *Journal of the Zoological Society of London*, **258**, 91-102

Samek, V. (1973) *Regiones fitogeográficas de Cuba*. Academia de Ciencias de Cuba.

Scherer, J.A. & Wilkinson, G.S. (1993) Evening bat isolation calls provide evidence for heritable signatures. *Animal Behaviour*, **46**, 847-860

Schilthuizen, M. (2000) Dualism and conflicts in understanding speciation. *BioEssays*, **22**, 1134-1140

Schmidt, S. (2005) Understanding wildlife: Behavioural ecology and acoustic communication in a Sri Lankan bat, Megaderma lyra. *Science in Sri Lanka*, **6**, 17-19

Schmieder, D.A., Kingston, T., Hashim, R. & Siemers, B.M. (2010) Breaking the trade-off: Rainforest bats maximize bandwidth and repetition rate of echolocation calls as they approach prey. *Biology Letters*, -

Schneider, S., Roessli, D. & Excoffier, L. (2000) Arlequin: A software for population genetics data analysis. In. Genetics and Biometry Lab, Dept. of Anthropology, University of Geneva., Geneva

Schnitzler, H.-U. & Kalko, E.K.V. (2001) Echolocation by insect-eating bats *BioScience*, **51**, 557-569

Schnitzler, H.-U., Moss, C.F. & Denzinger, A. (2003) From spatial orientation to food acquisition in echolocationg bats. *Trends in Ecology and Evolution*, **18**, 386-394

Siemers, B.M. & Schnitzler, H.-U. (2004) Echolocation signals reflect niche differentiation in five sympatric congeneric bat species *Nature*, **429**, 657-661

Silva Taboada, G. (1979) *Los murcielagos de Cuba*, 1 edn. Editorial Academia, La Habana, Cuba.
Simmons, N.B. (2005) Order Chiroptera. *Mammal species of the world: A taxonomic and geographic reference* (ed. by D.E. Wilson and D.M. Reeder), pp. 312-529. Bucknell University.
Smith, M.F. & Patton, J.L. (1991) Variation in mitochondrial Cytochrome b sequence in natural populations of South American Akodontine rodents (Muridae: Sigmodontinae). *Molecular Biology and Evolution*, **8**, 85-103
Straney, D.O., Smith, M.H., Greenbaum, I.F. & Baker, R.J. (1979) Biochemical genetics. *Biology of bats of the New World family Phyllostomatidae, part iii* (ed. by R.J. Baker, J.K. Jones and D.C. Carter), pp. 157-176. Special Publications Museum Texas Tech University.
Surlykke, A., Futtrup, V. & Tougaard, J. (2003) Prey-capture success revealed by echolocation signals in pipistrelle bats (Pipistrellus pygmaeus). *Journal of Experimental Biology*, **206**, 93-104
Swofford, D.L. (2003) Paup*. Phylogenetic analysis using parsimony (*and other methods). In. Sinauer Associates, Sunderland, Massachusetts
Sztencel, J., Onka, A., Jones, G., Bogdanowicz, W. & Aw (2009) Skull morphology of two cryptic bat species: Pipistrellus pipistrellus and P. pygmaeus a 3d geometric morphometrics approach with landmark reconstruction. *Acta Chiropterologica*, **11**, 113-126
Tamura, K., Dudley, J., Nei, M. & Kumar, S. (2007a) Mega4: Molecular evolutionary genetics analysis (Mega) software version 4.0. *Molecular Biology and Evolution*, **24**, 1596-1599
Tamura, K., Dudley, J., Nei, M. & Kumar, S. (2007b) Molecular evolutionary genetics analysis (Mega). In. Molecular Biology and Evolution 24:1596-1599
Teacher, A.G.F. & Griffiths, D.J. (2011) Hapstar: Automated haplotype network layout and visualization. *Molecular Ecology Resources*, **11**, 151-153
Teeling, E.C., Springer, M.S., Madsen, O., Bates, P., O'Brien, S.J. & Murphy, W.J. (2005) A molecular phylogeny for bats illumunates biogeography and the fossil records. *Science*, **307**, 580-887
Van Den Bussche, R.A. (1992) Restriction-site variation and molecular systematics of New World Leaf-nosed bats. *Journal of Mammalogy*, **73**, 29-42
Van Den Bussche, R.A. & Baker, R.J. (1993b) Molecular phylogenetics of the New World bat genus Phyllostomus based on Cytochrome b DNA sequence variation. *Journal of Mammalogy*, **74**, 793-802
Von Helversen, D. & Von Helversen, O. (2003) Object recognition by echolocation: A nectar-feeding bat exploiting the flowers of a rain forest vine. *Journal of Comparative Physiology A. Sensory, Neural, and Behavioral Physiology*, **189**, 327-336
Weir, B.S. (1996) *Genetic data analysis II*. Sinauer, Sunderland, MA.
Weir, B.S. & Cockerham, C.C. (1984) Estimating F-statistics for the analysis of population structure. *Evolution and Human Behavior*, **38**, 1358-1370
Wilkinson, G.S. & Chapman, A.M. (1991) Length and sequence variation in Evening bat d-loop mtDNA. *Genetics*, **128**, 607-617
Willig, M.R., Presley, S.J., Bloch, C.P. & Genoways, H.H. (2009) Macroecology of Caribbean bats: Effects of area, elevation, latitude, and hurricane-induced disturbance.

Evolution, ecology, and conservation of island bats (ed. by T. Fleming and P. Racey), pp. 216-264. University of Chicago Press, Chicago.

Zöfel, P. (1988) *Statistik in der Praxis*. Gustav Fischer Verlag, Stuttgart.

i want morebooks!

Buy your books fast and straightforward online - at one of world's fastest growing online book stores! Environmentally sound due to Print-on-Demand technologies.

Buy your books online at
www.get-morebooks.com

Kaufen Sie Ihre Bücher schnell und unkompliziert online – auf einer der am schnellsten wachsenden Buchhandelsplattformen weltweit! Dank Print-On-Demand umwelt- und ressourcenschonend produziert.

Bücher schneller online kaufen
www.morebooks.de

VDM Verlagsservicegesellschaft mbH
Heinrich-Böcking-Str. 6-8
D - 66121 Saarbrücken

Telefon: +49 681 3720 174
Telefax: +49 681 3720 1749

info@vdm-vsg.de
www.vdm-vsg.de

Printed by Books on Demand GmbH, Norderstedt / Germany